FORGOTTEN
MURFREESBORO
TENNESSEE

IF WALLS COULD TALK

EROS EMERSON RUSSELL

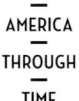

AMERICA
THROUGH
TIME

America Through Time
www.through-time.com
office@through-time.com

First published 2025
Copyright © Eros Emerson Russell 2025

ISBN 978-1-63499-509-2

Typeset in Trade Gothic 10pt on 15pt
Printed and bound in England

CONTENTS

INTRODUCTION

Fascinated by the forgotten and abandoned buildings, homes, and miscellaneous belongings left behind that are hidden beneath the backroads and desolate small towns of Middle Tennessee, I fell in love with driving down backroads or just driving with no destination when I first started driving as a teenager. There was not much to do where I grew up but drive around town, but I quickly fell in love with the excitement of finding the remnants of what had been left behind from those who came (and left) before us. Some left everything behind, and I think being able to capture history or pieces of someone's life they left behind and the mystery that surrounds all of that through photography is what truly fascinated me.

Driving aimlessly through towns that were once war communities and served as pillars of other historical times and events, as well as abandoned farmland, there is a specific feeling you get driving down these backroads where dilapidated homes stand on their last leg; you wonder who used to walk up those hand-built wooden steps, listen to the rain hit that rusty tin roof or what that house with the white picket fence looked like before it was engulfed in vines. You wonder what memories took place in the front yard that is now a forest of overgrown grass and weeds.

Using photography as a way to bring these places and things to life, urban exploring is more than just taking pictures of abandoned places; it is taking pictures of places and things that have been forgotten and left behind and discovering the story it tells without words. It is finding beauty in rusted tin roofs, planks of wood hanging on by the last splinter of splitting wood, missing floorboards, dust and grime, the random patterns of paint that align houses, and pieces of woodwork where you can tell at one point, decades or centuries ago, it was freshly painted but you can see the original wood where age, rain, and nature has taken its course.

It is jerking your steering wheel to pull over quickly because you are in awe of the remains of a story you are unable to hear; it is seeing a legacy and lesson where some people see a rotting structure. It is finding art in history and beauty in the forgotten.

This book captures some of the forgotten homes, lives, and history of Murfreesboro, Tennessee, and surrounding communities. Most of these pictures were taken in Murfreesboro as well as surrounding unincorporated communities such as Rockvale, Christiana, Unionville, Readyville, Woodbury, and more. While these photos are only a portion of the urbex photography I have done over the years, I am hopeful that, as our community is rapidly growing, these images will allow the history of our communities to be documented, remembered, and honored.

My only goal in anywhere I explore and photograph is to bring life to what others may see as ruins, to capture history through a lens that is capable of seeing beauty in nature, reclaiming what is left behind, and I will always follow and respect the motto "take only photographs and leave only footprints."

1

NOBODY'S HOME

In the quiet corners of forgotten landscapes and the fringes of Middle Tennessee lies a haunting reminder of lives once lived. These abandoned homes, frozen in time, stand as silent witnesses to the passage of years, their walls echoing with stories untold and memories left behind. Each photograph in this collection captures more than just decay; it encapsulates the beauty and mystery of abandonment.

Through the lens, we are invited to explore these homes, where families once resided, children once played, and where memories were made. Now, nature reclaims its dominion with creeping ivy and shattered windows frame scenes frozen in time. Each image tells a tale of departure—whether abrupt or gradual—leaving interiors adorned with remnants of lives interrupted: a child's toy forgotten on a dusty floor, a weathered photo album opened to faded smiles, or a dining table set for a meal never shared.

The allure of these abandoned homes lies not only in their dilapidation, but in the sense of mystery they evoke. Who were the inhabitants who once called these places home? What dreams and tragedies unfolded within these walls, now surrendered to the ravages of neglect?

As we embark on this visual journey, let us contemplate the fragile boundary between presence and absence, and the enduring resonance of places left behind. Through the lens of abandonment, we uncover not only the decay of bricks and mortar but also the enduring essence of human existence—a testament to resilience, impermanence, and the quiet persistence of memory.

The Untold Story of a Forgotten Farmhouse

This abandoned Rockvale home, hidden by overgrown land, still holds furniture, toys, and children's clothing from the 1970s and '80s, telling stories of the past lives that once unfolded here.

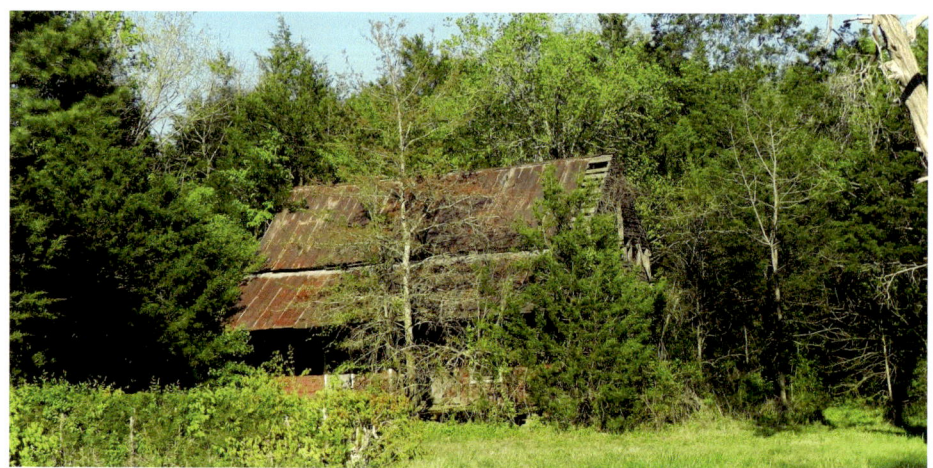

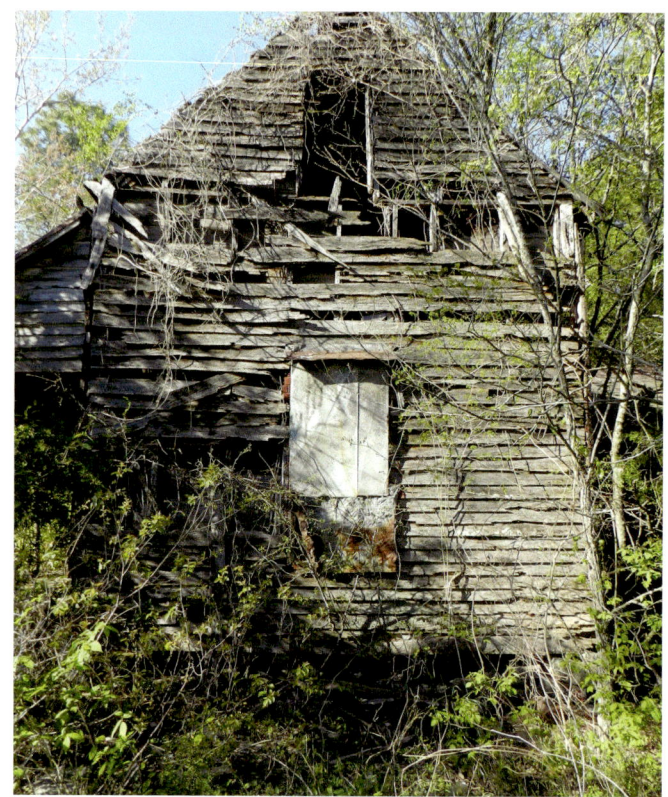

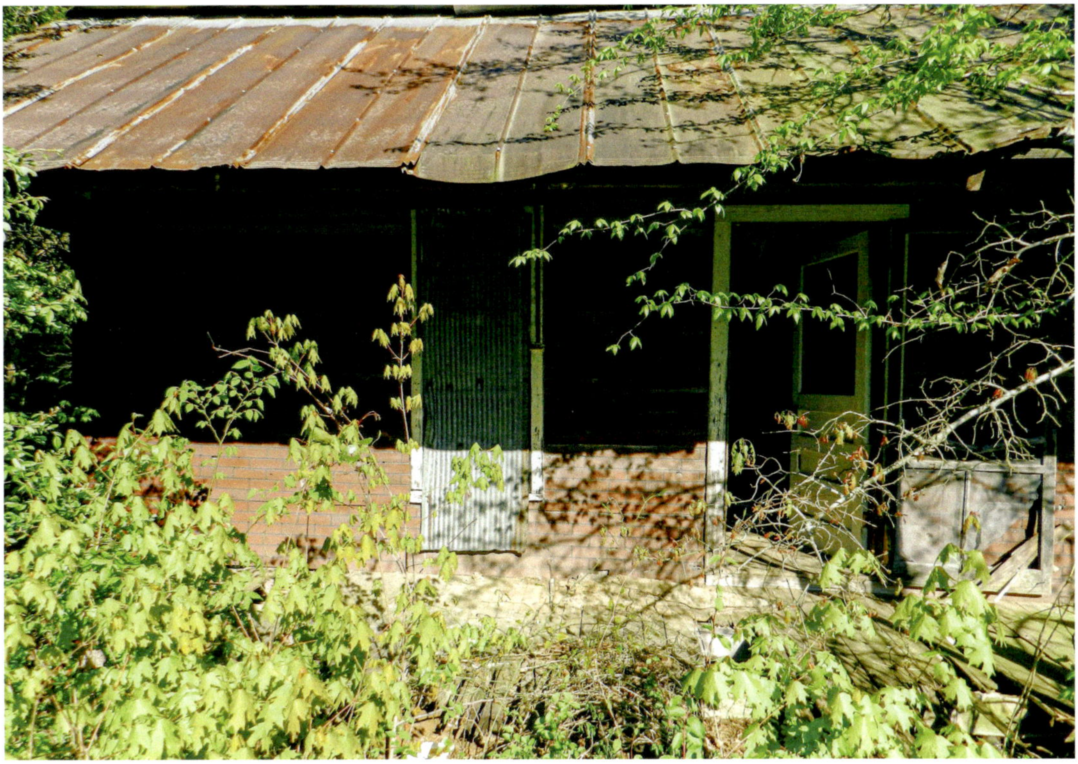

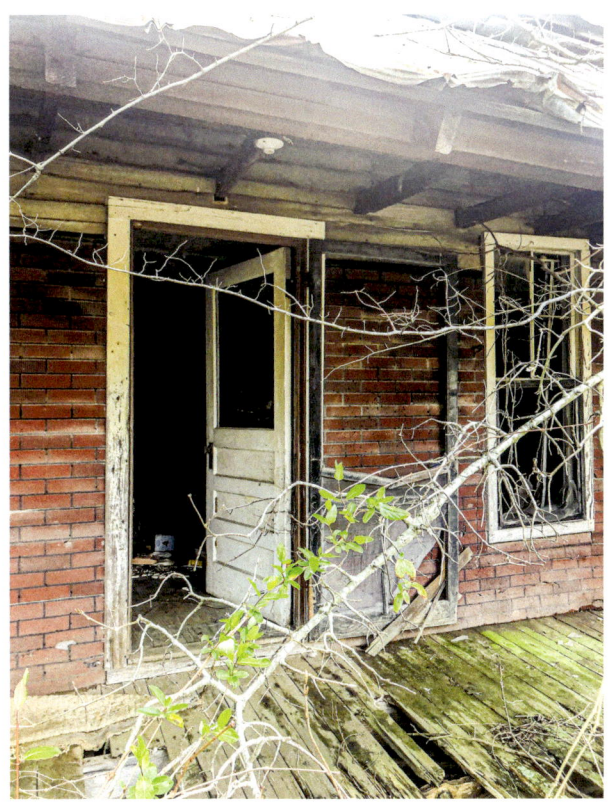

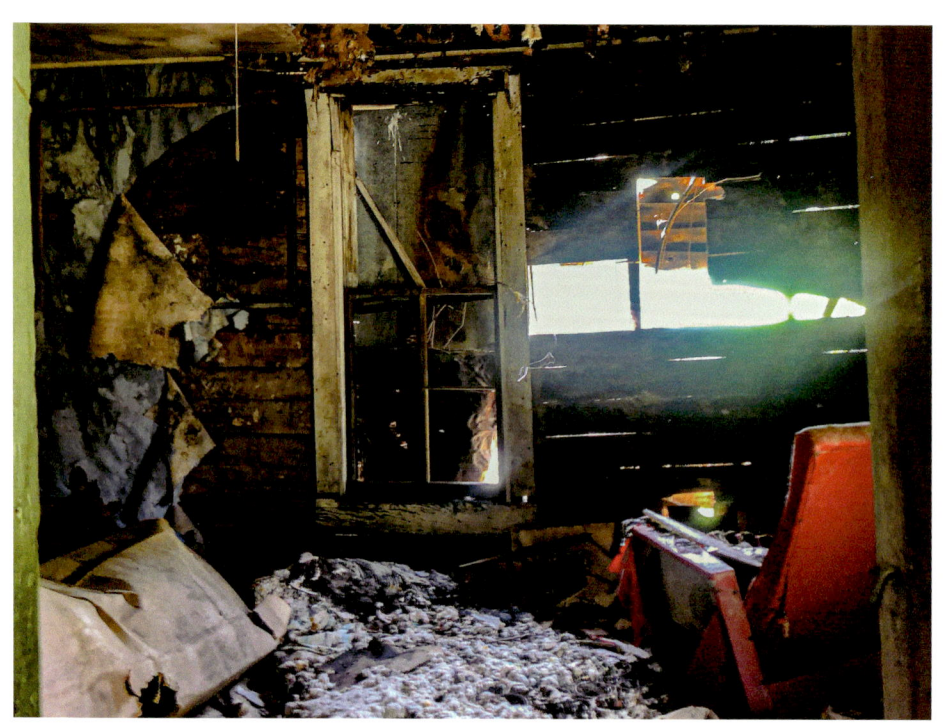

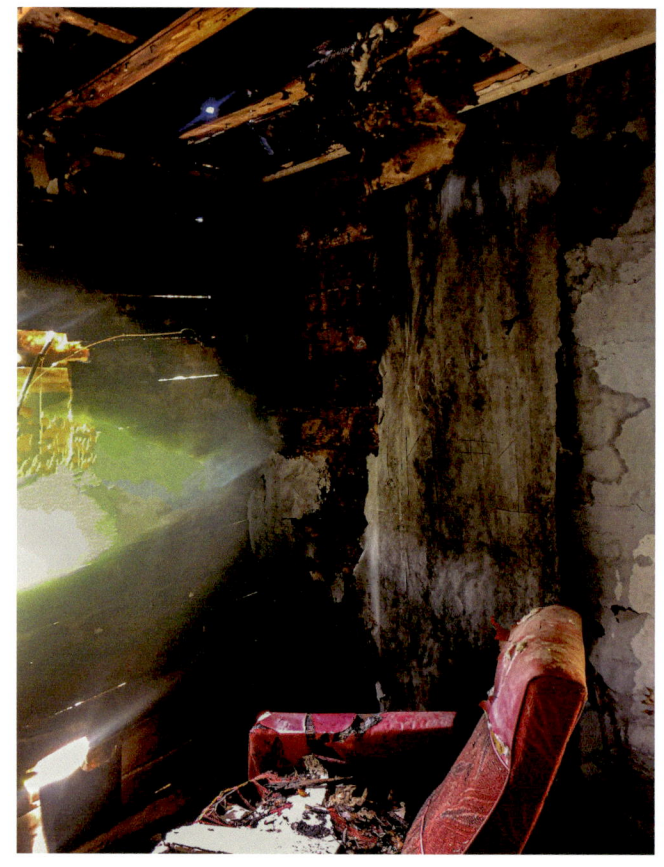

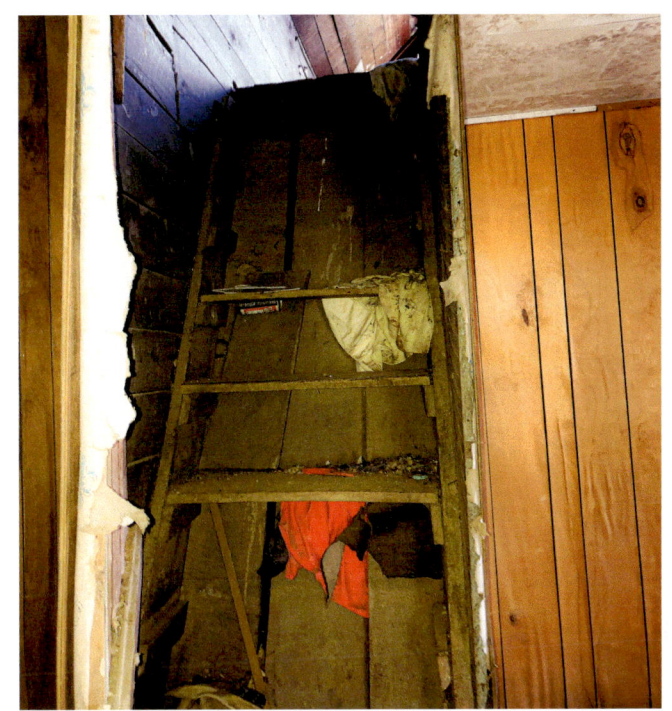

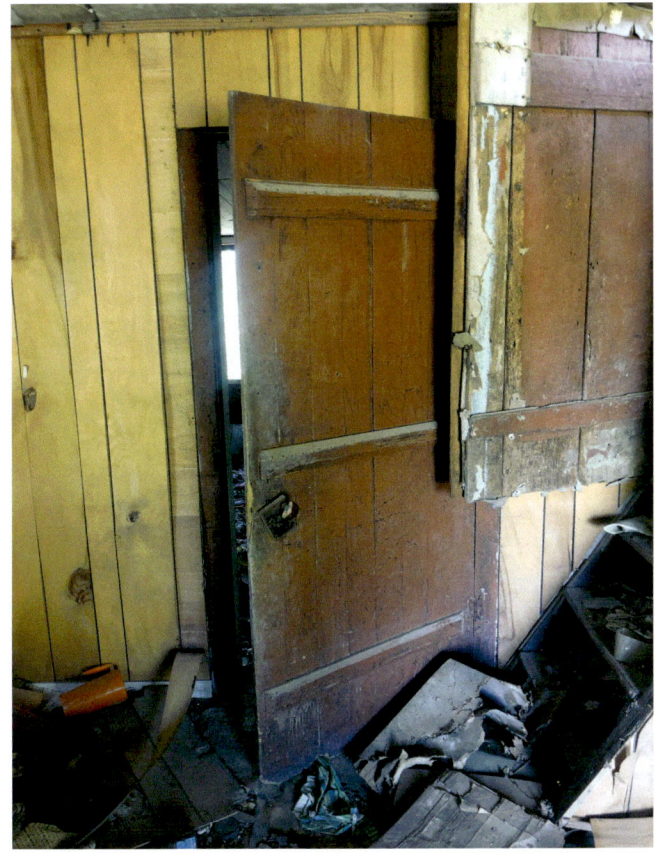

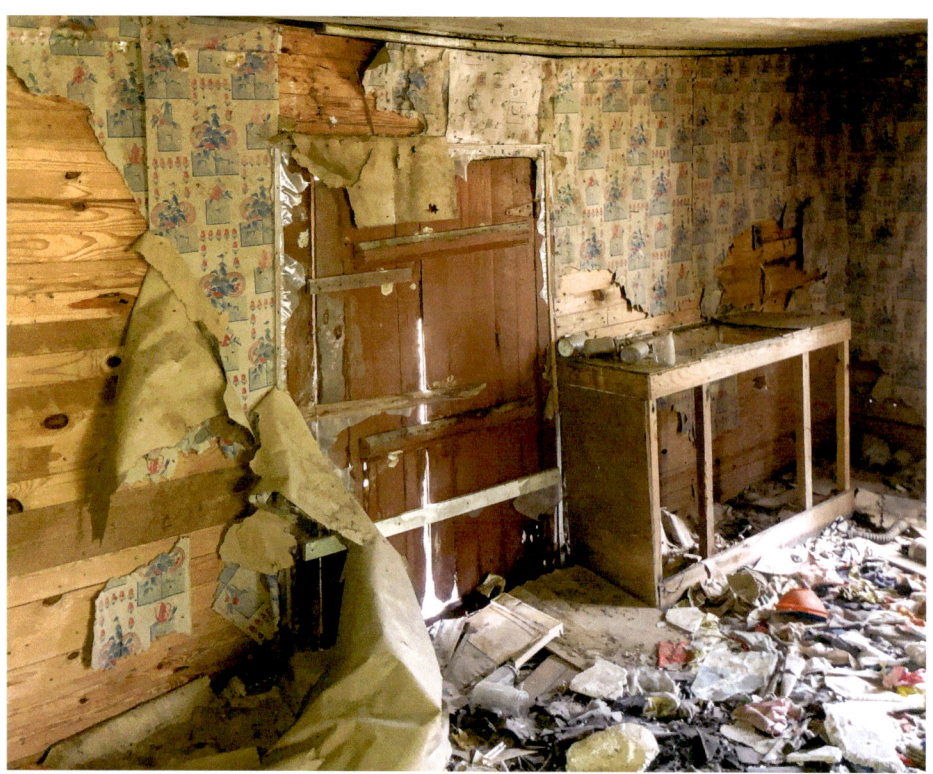

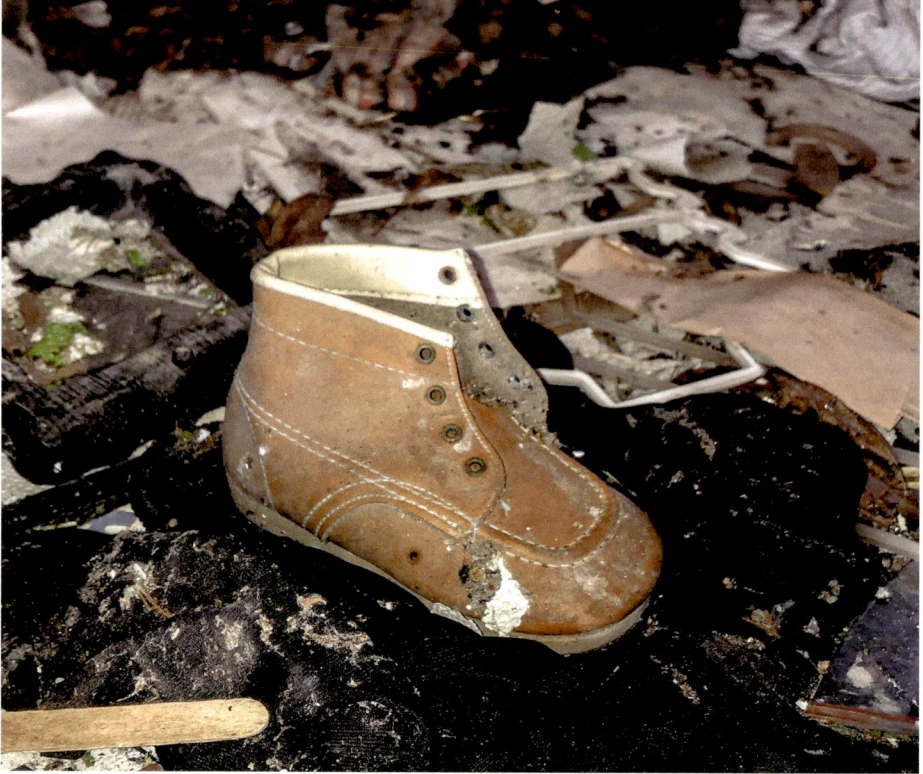

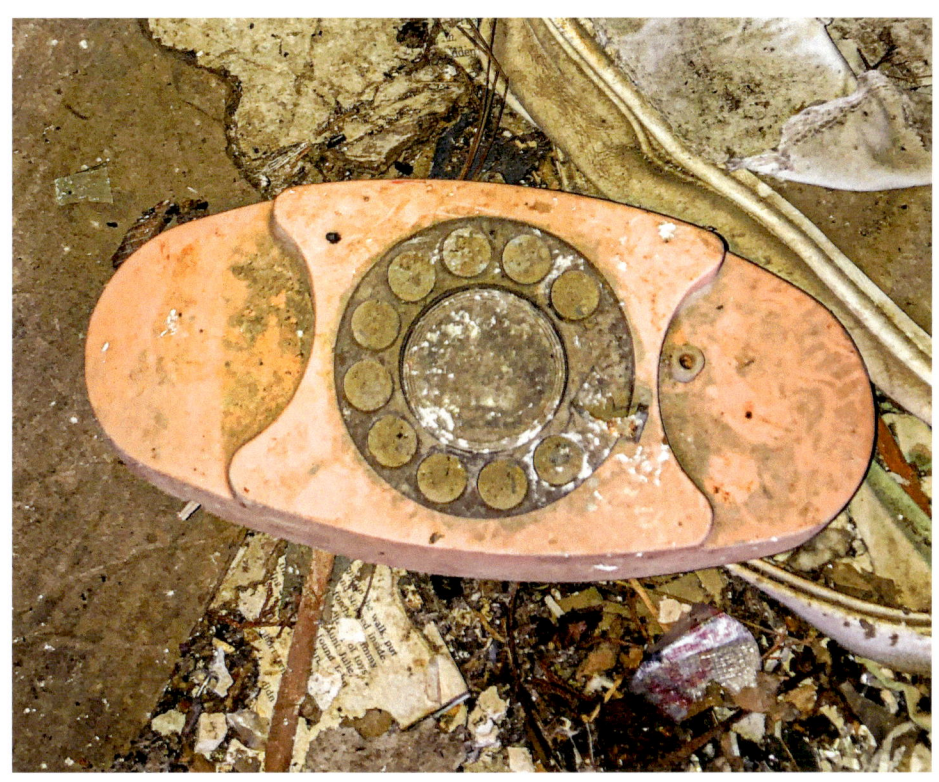

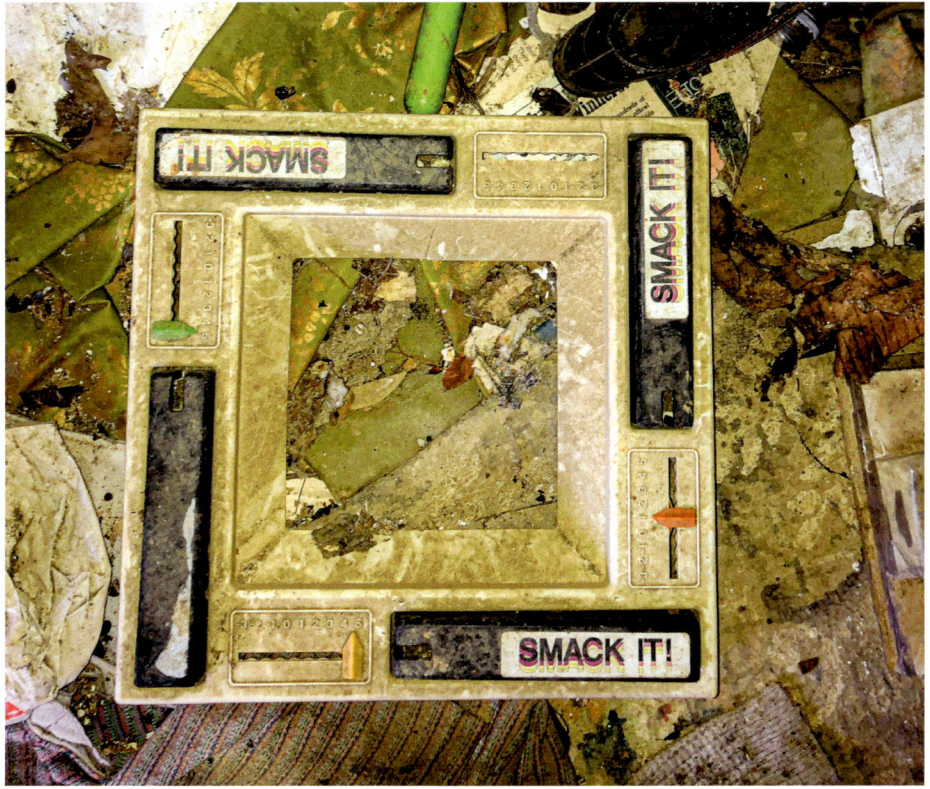

Where Avocado Green Meets Rose Quartz:
A Bygone Bathroom Dream

From the roadway, I could only see the rooftop peeking through vines, hiding years of neglect. Inside, personal artifacts like a pink bathroom, floral wallpaper, clothes still hanging in the closet, food left in the fridge, and a calendar from 1988 made me wonder who lived here and why they left so suddenly.

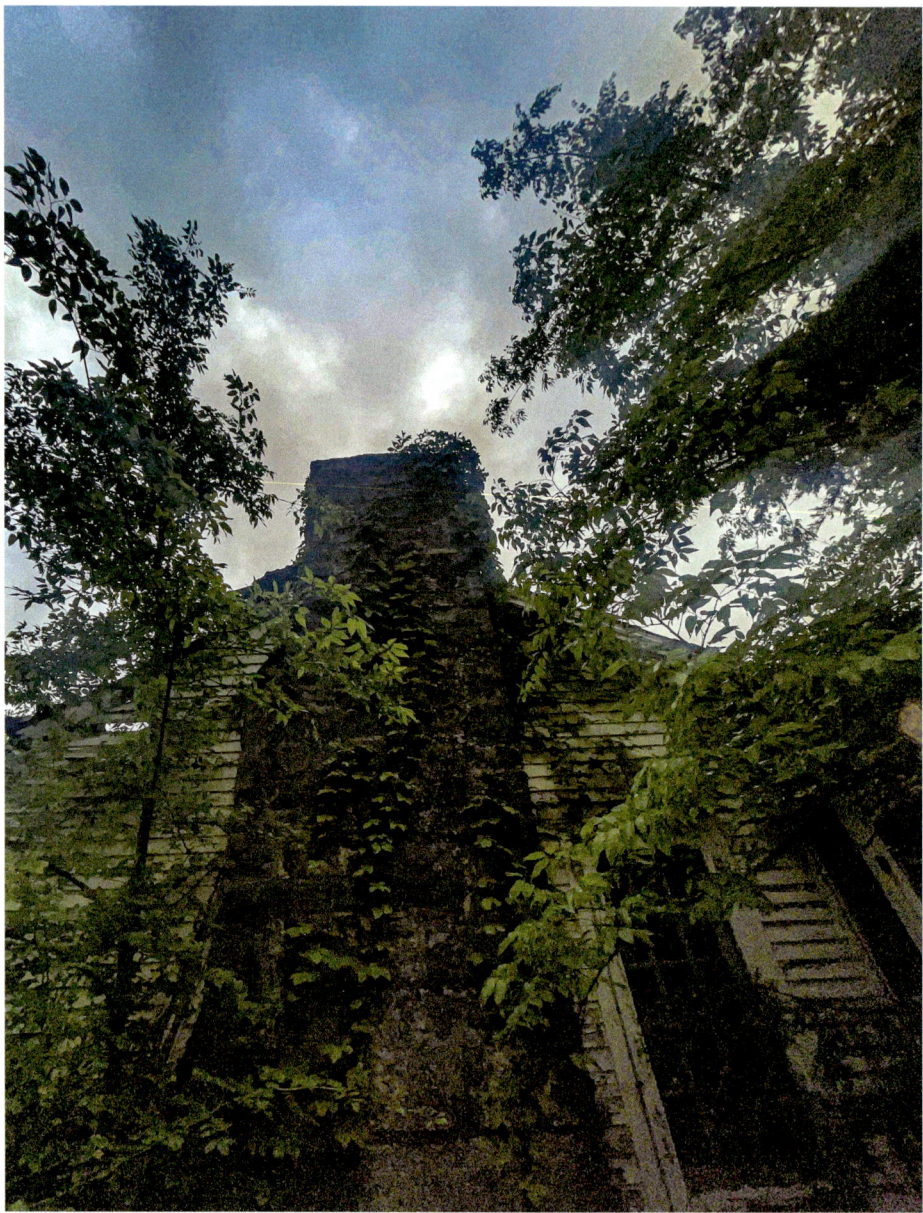

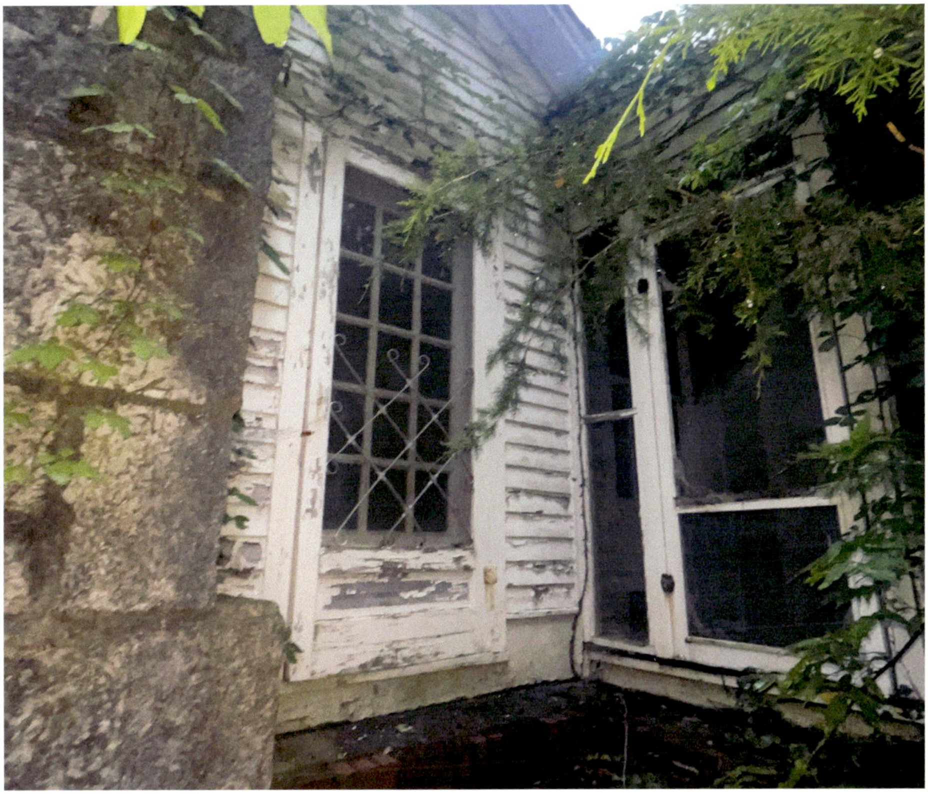

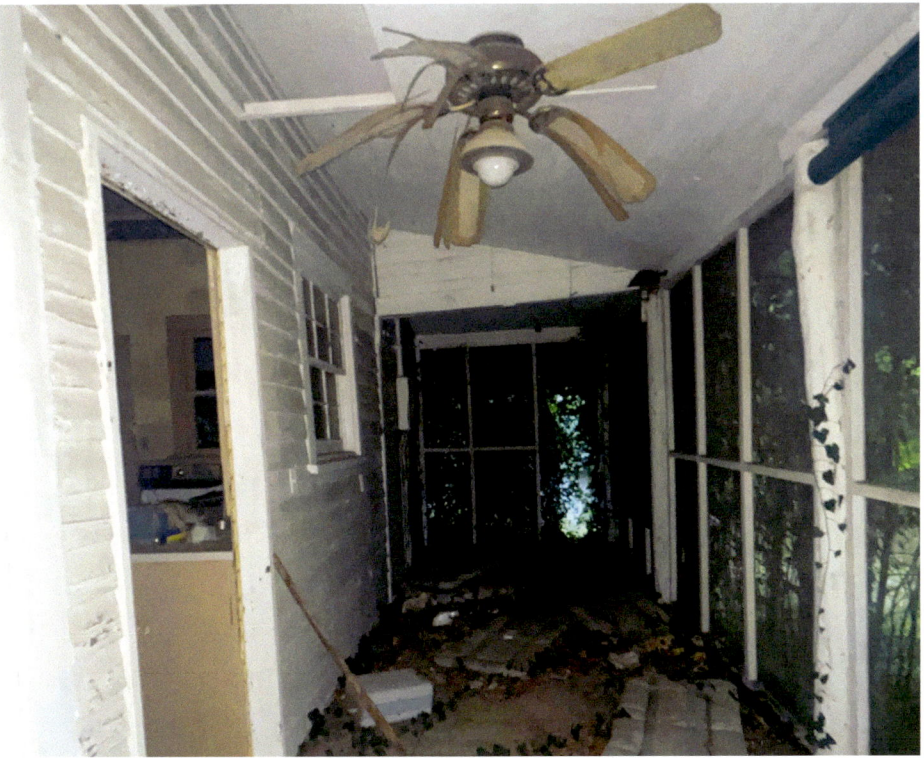

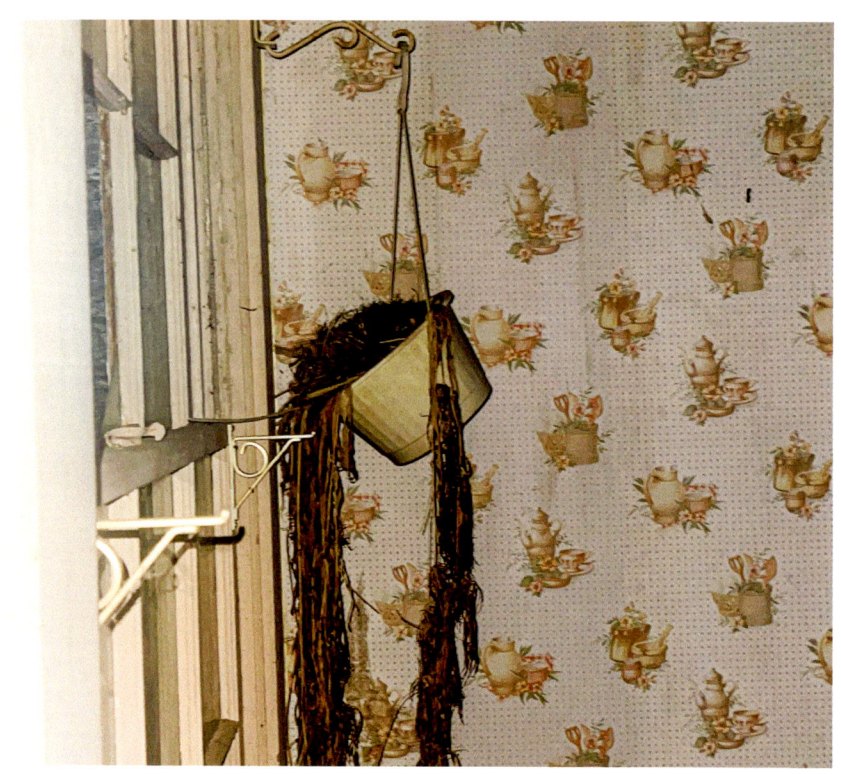

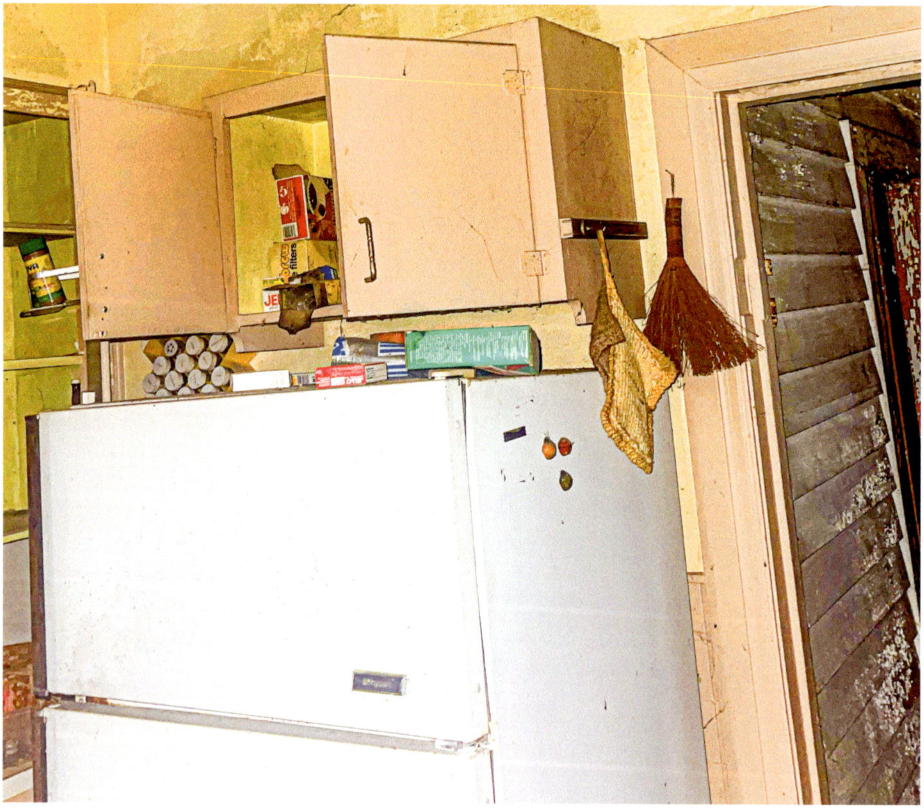

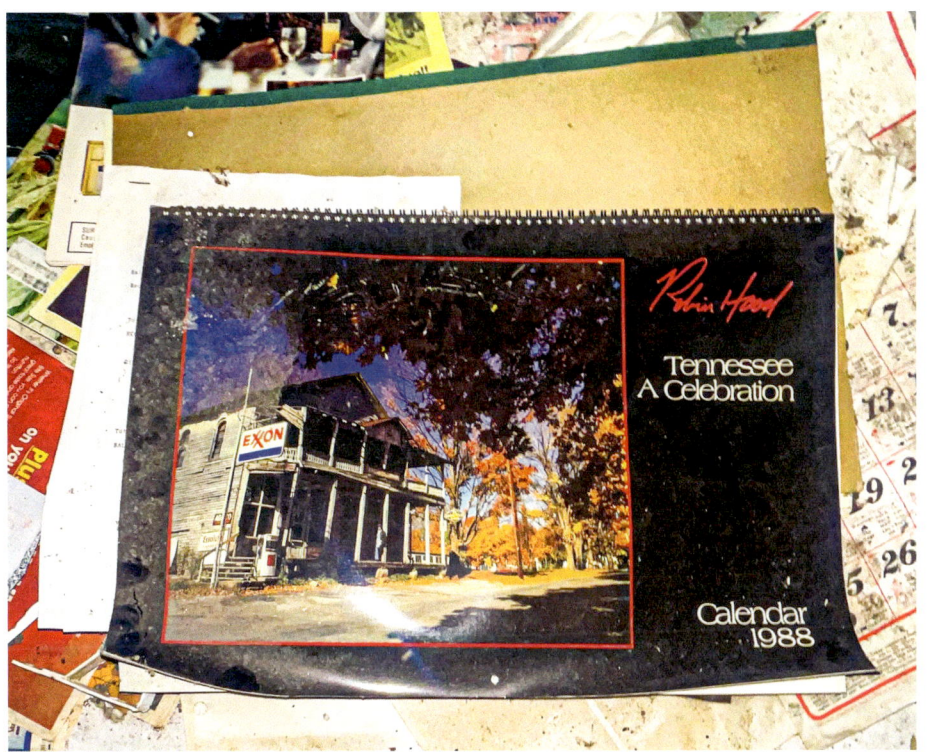

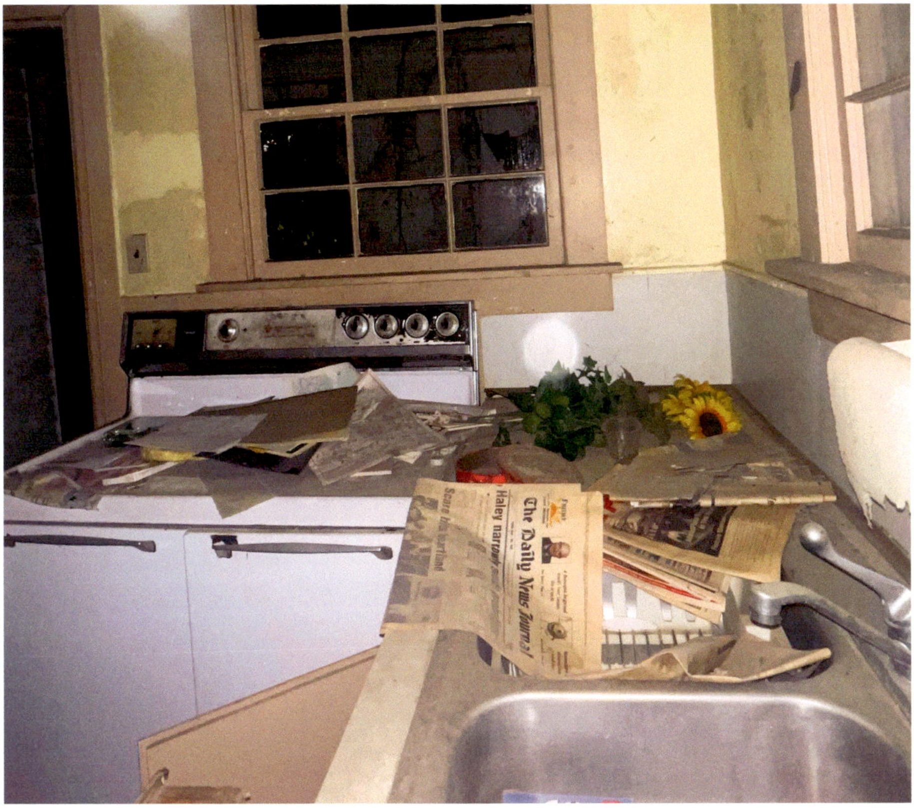

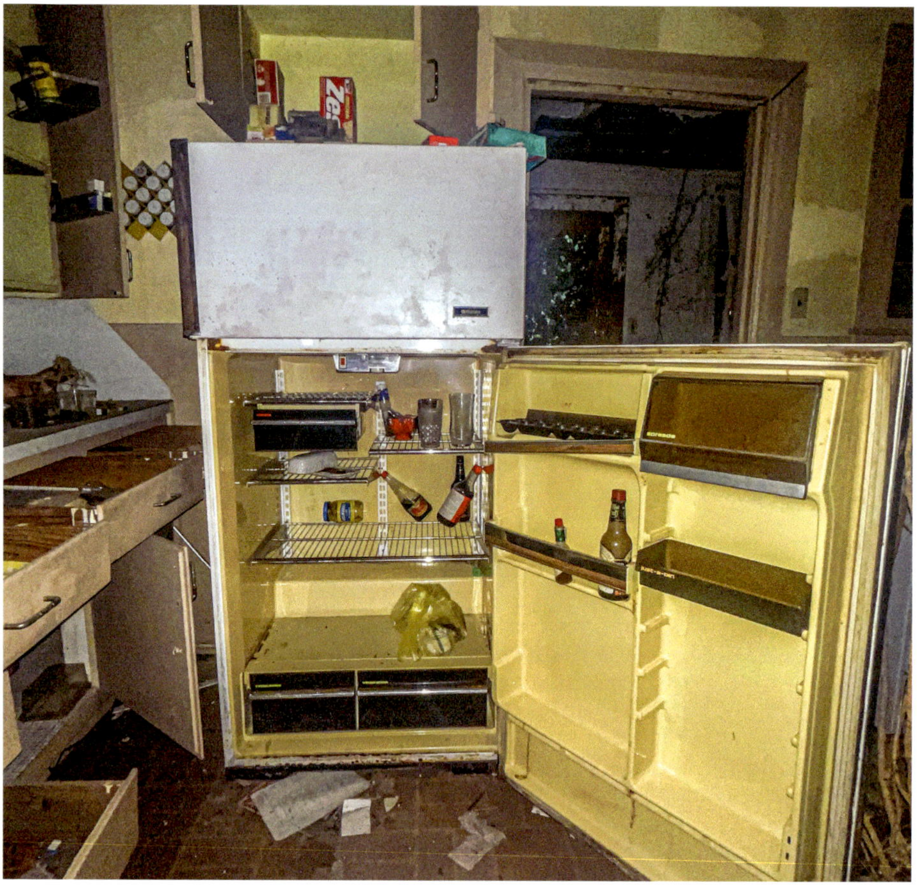

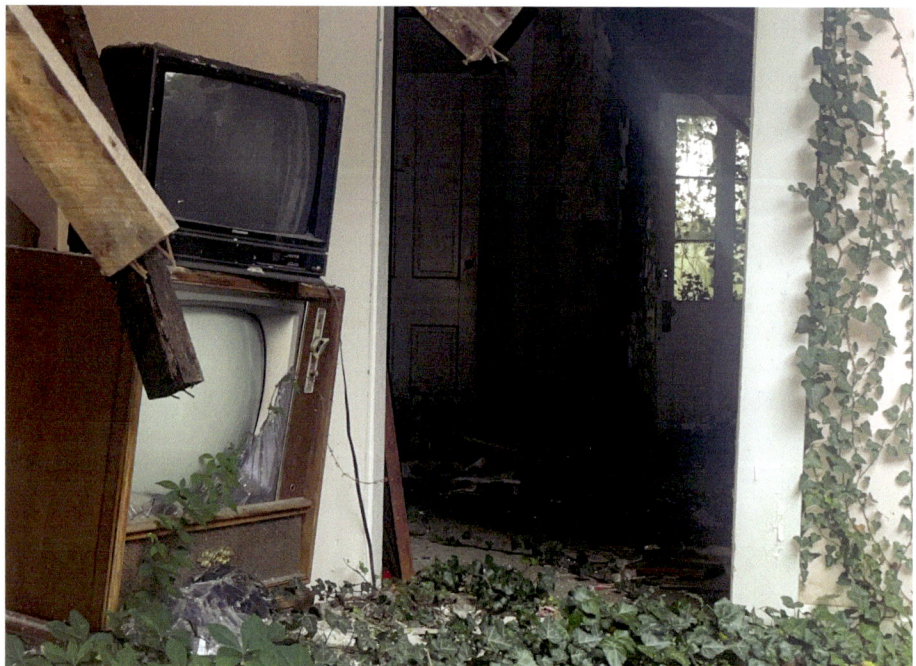

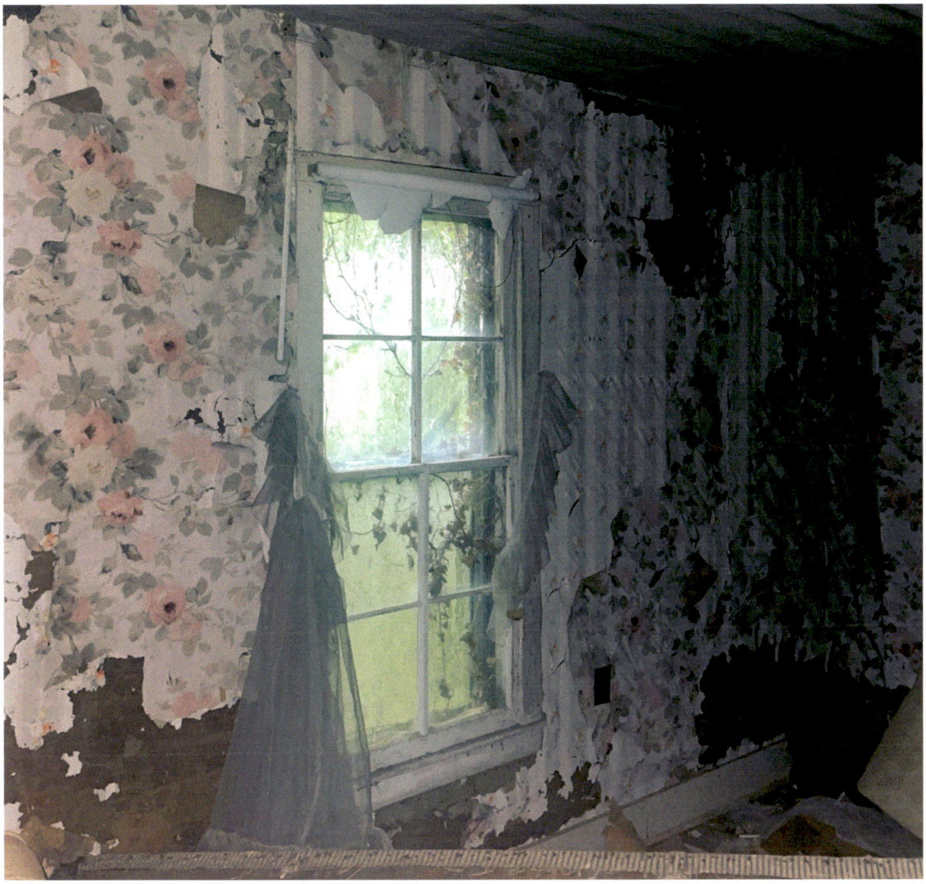

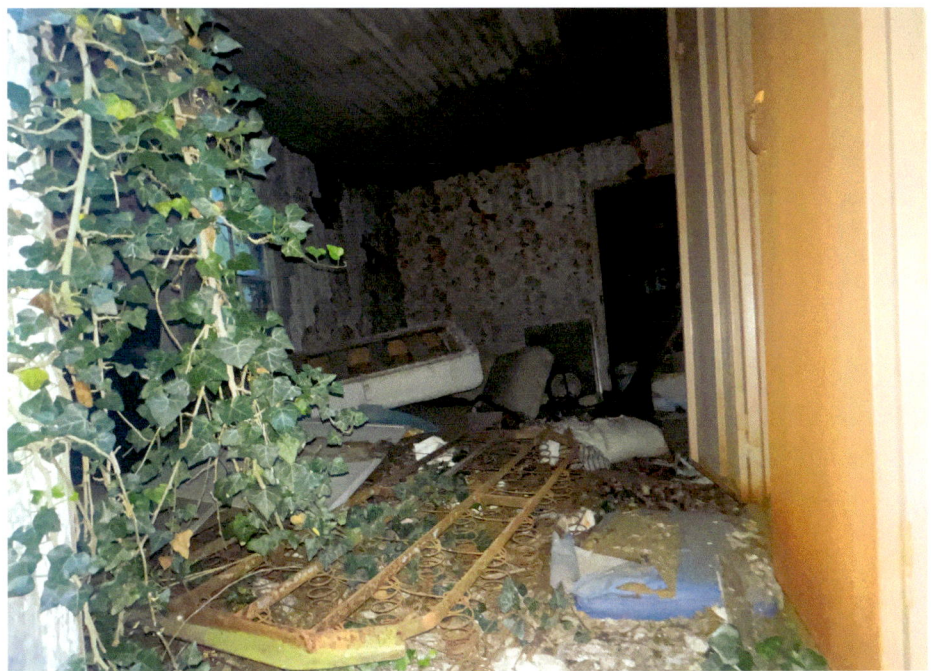

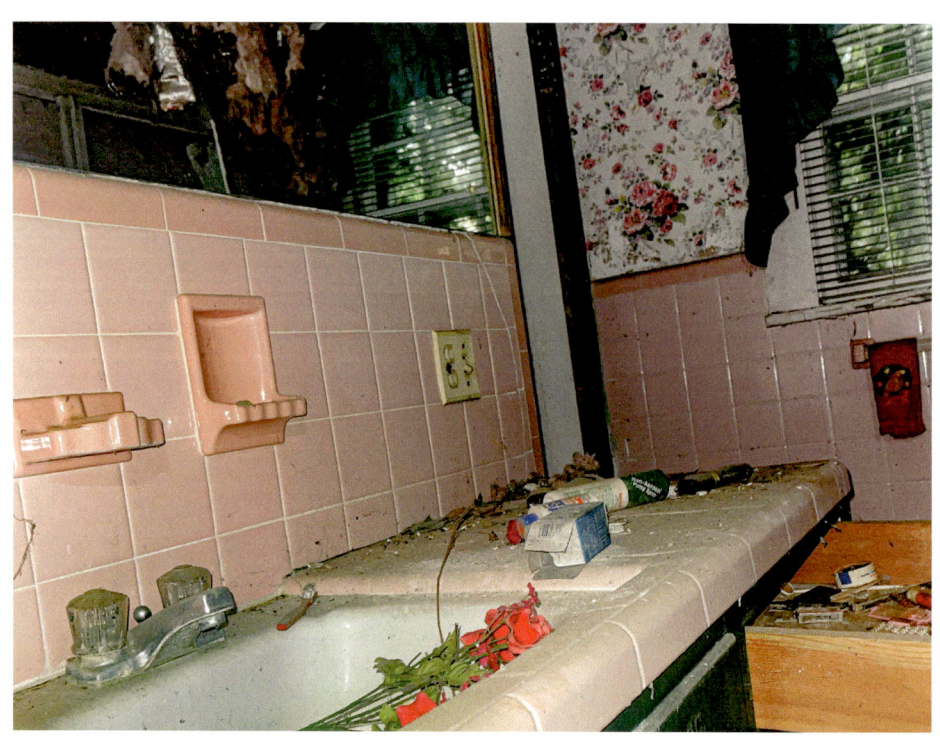

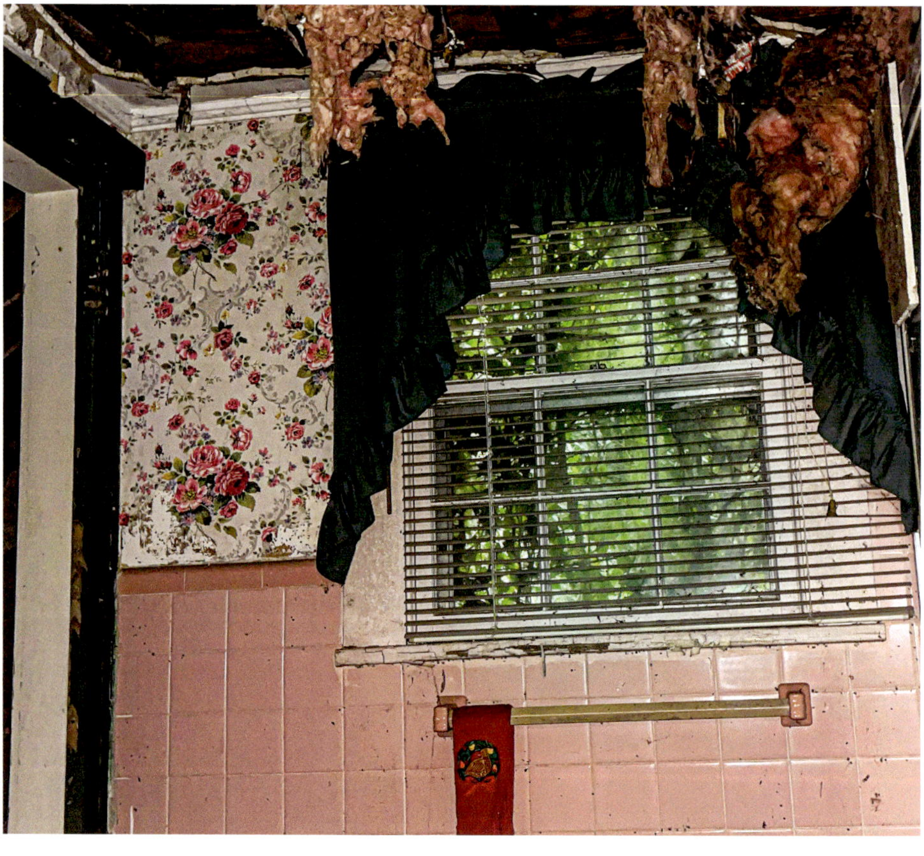

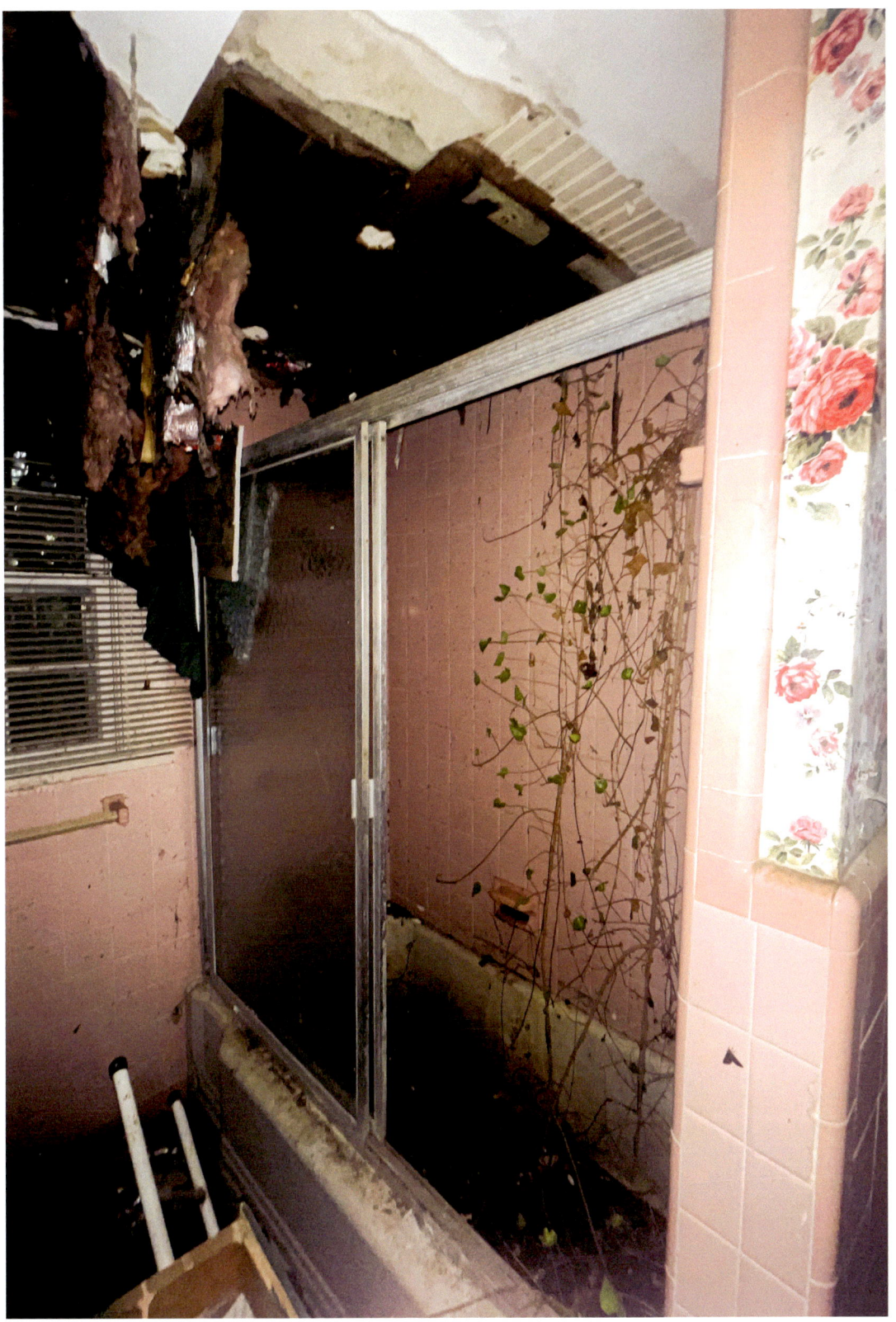

Weathered in Woodbury

This weathered Woodbury home, secluded along a winding road, exudes history and age. Sunlight filters through a broken window, casting a glow on framed memories and scattered trophies, reminiscent of silent ghosts from the past.

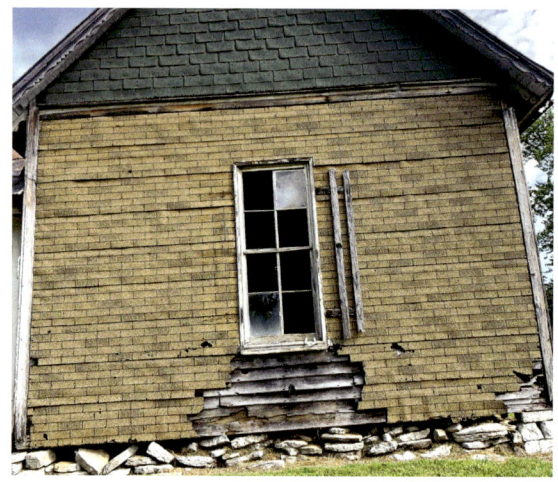

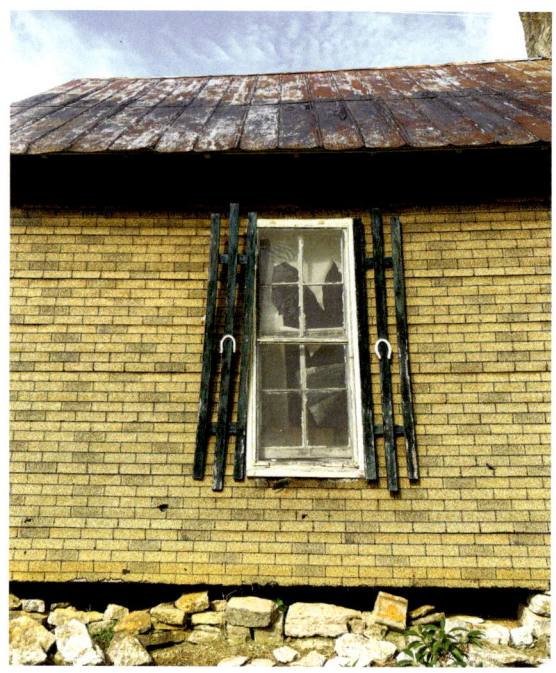

Beauty Finds a Foothold: Flowers Bloom in the Ruins of the Past

This Unionville home, hidden amid overgrown landscape, captivated me with its sunlit rusted tin roof. Inside, handcrafted cabinetry and shelving caught my eye, while sunlight peeked through the bowing lumber. Imagining the lives that once unfolded here, I felt like I could see the memories this home once shared with others.

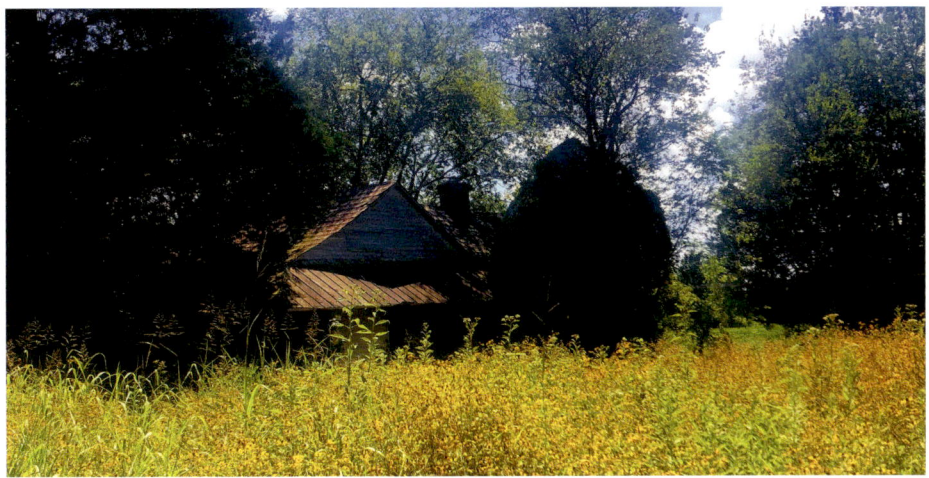

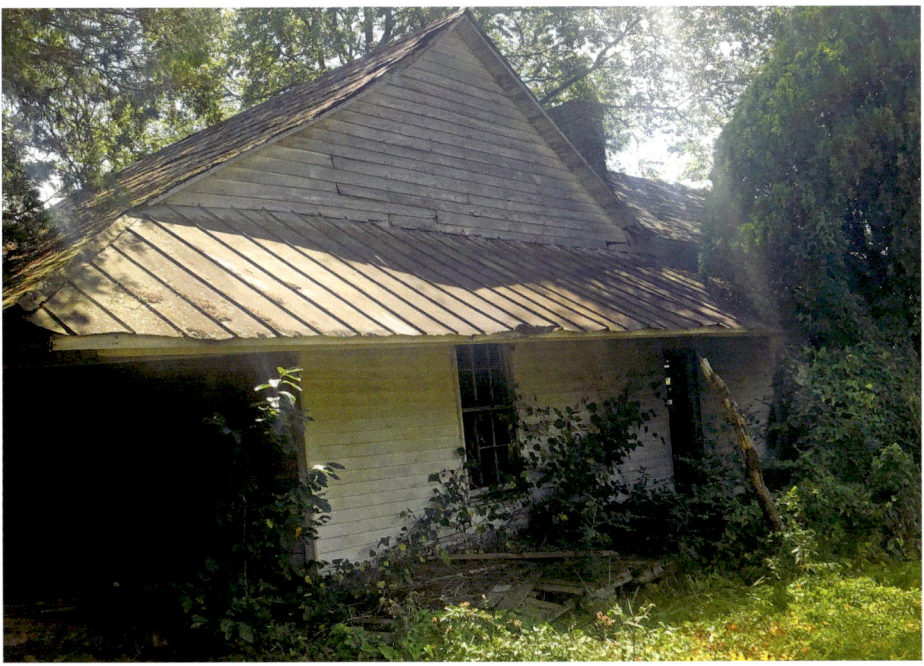

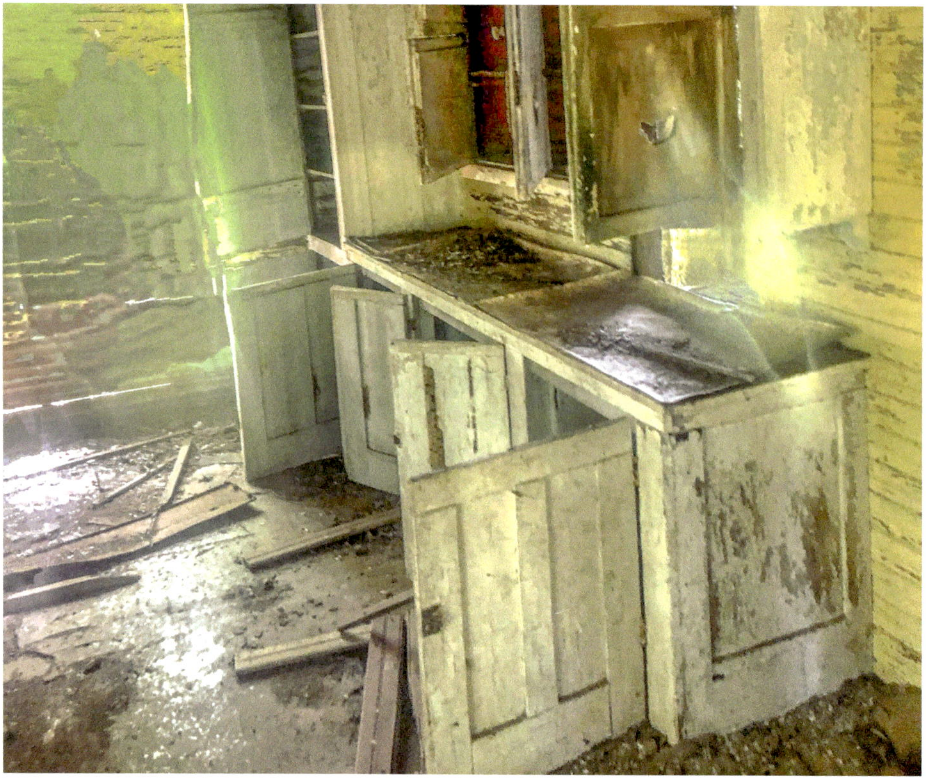

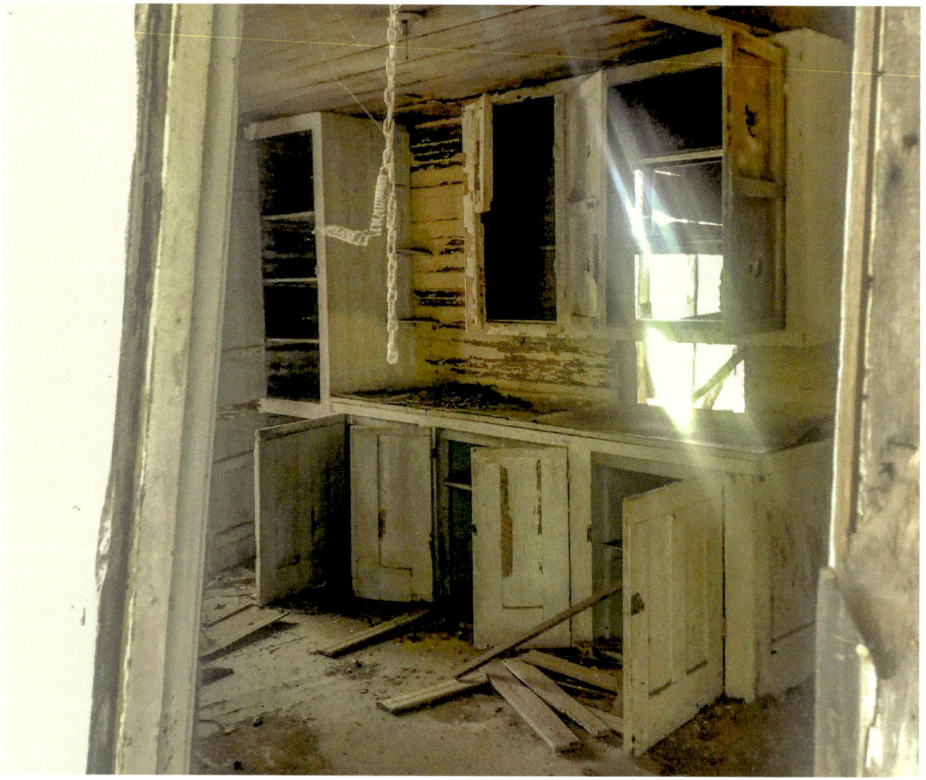

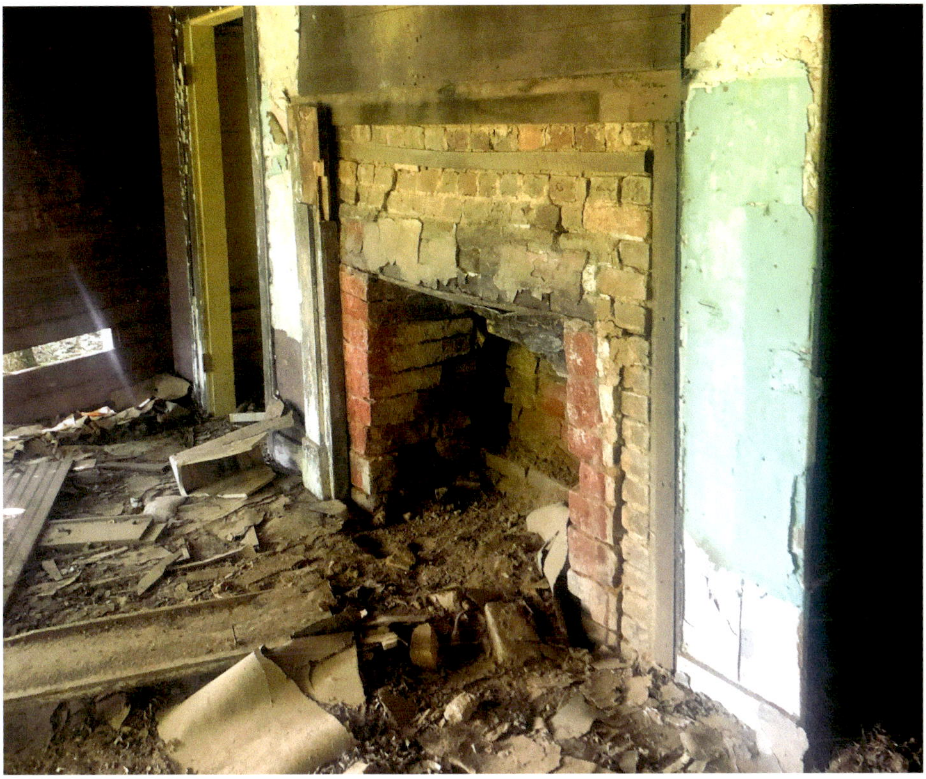

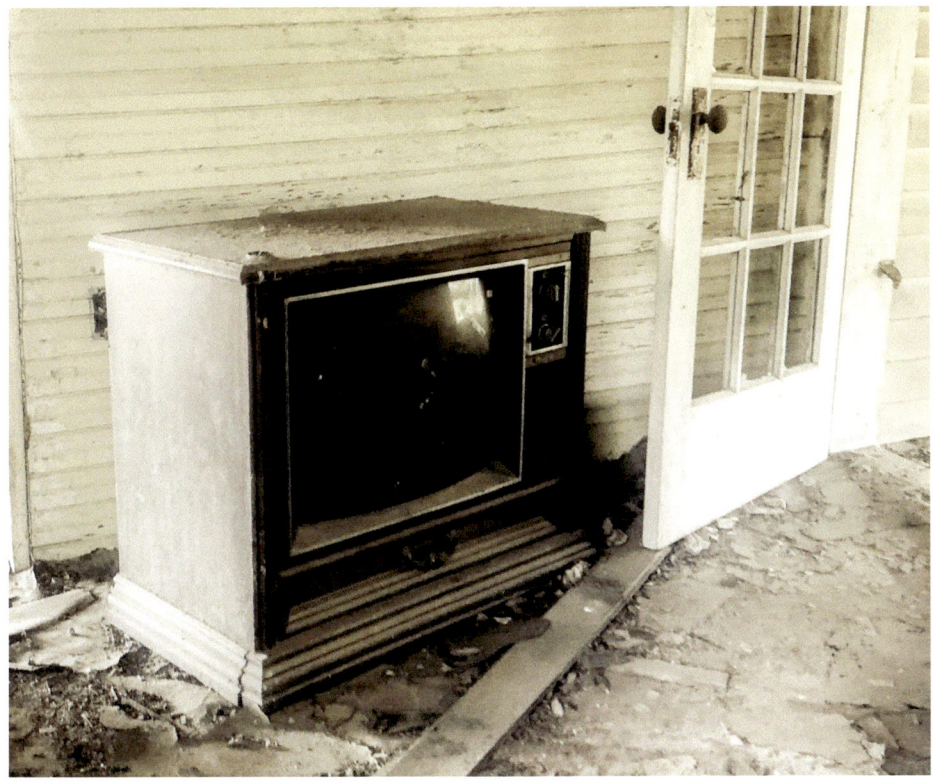

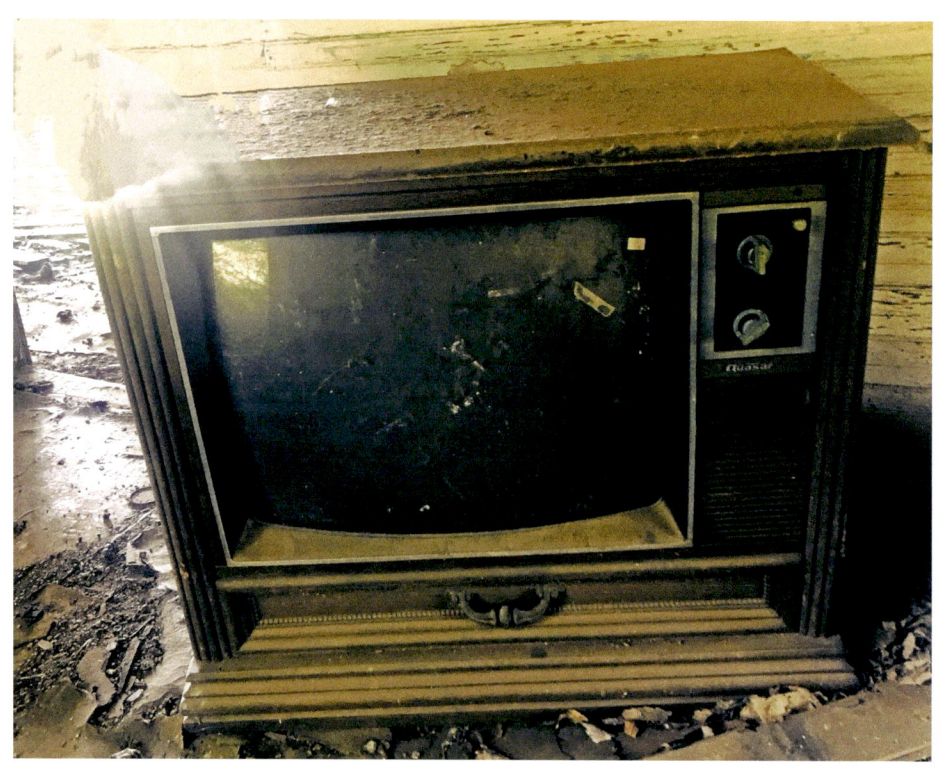

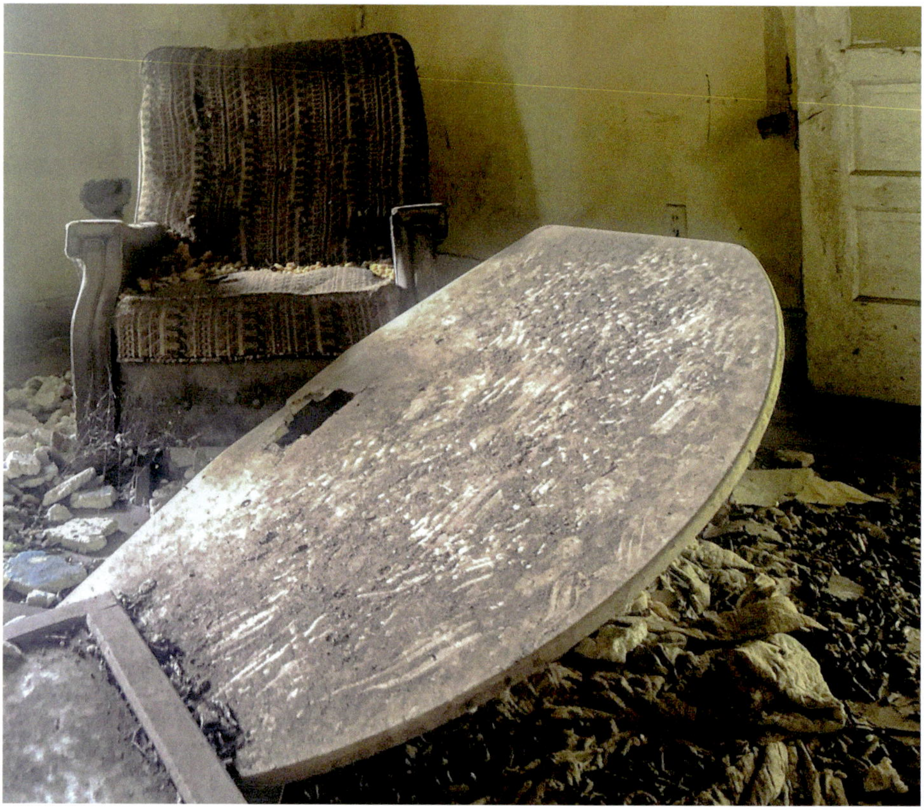

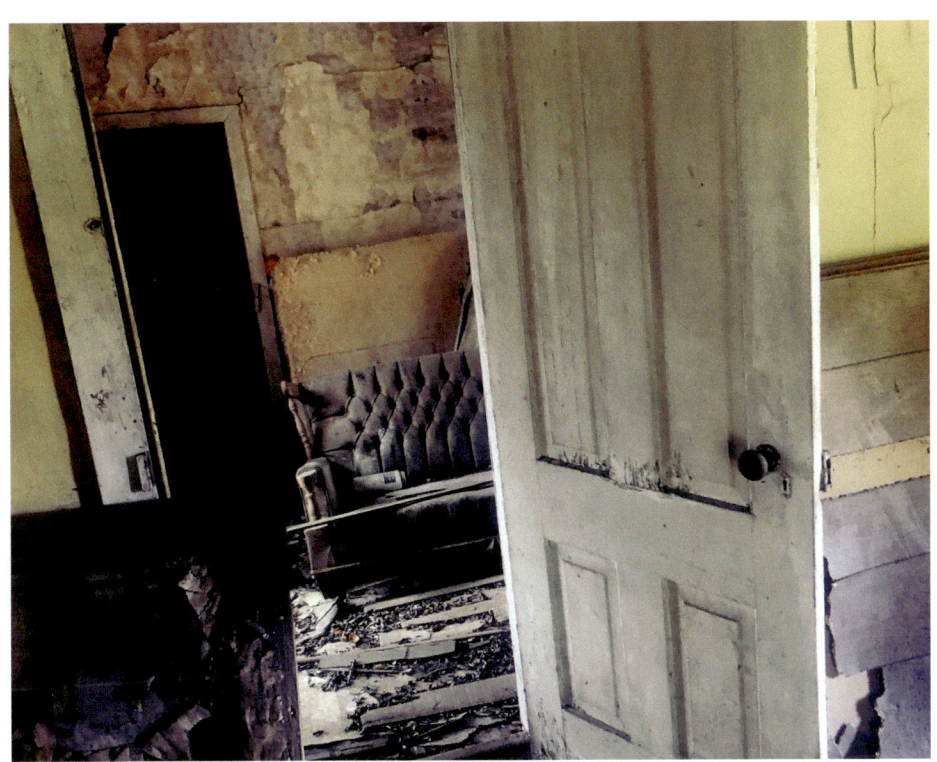

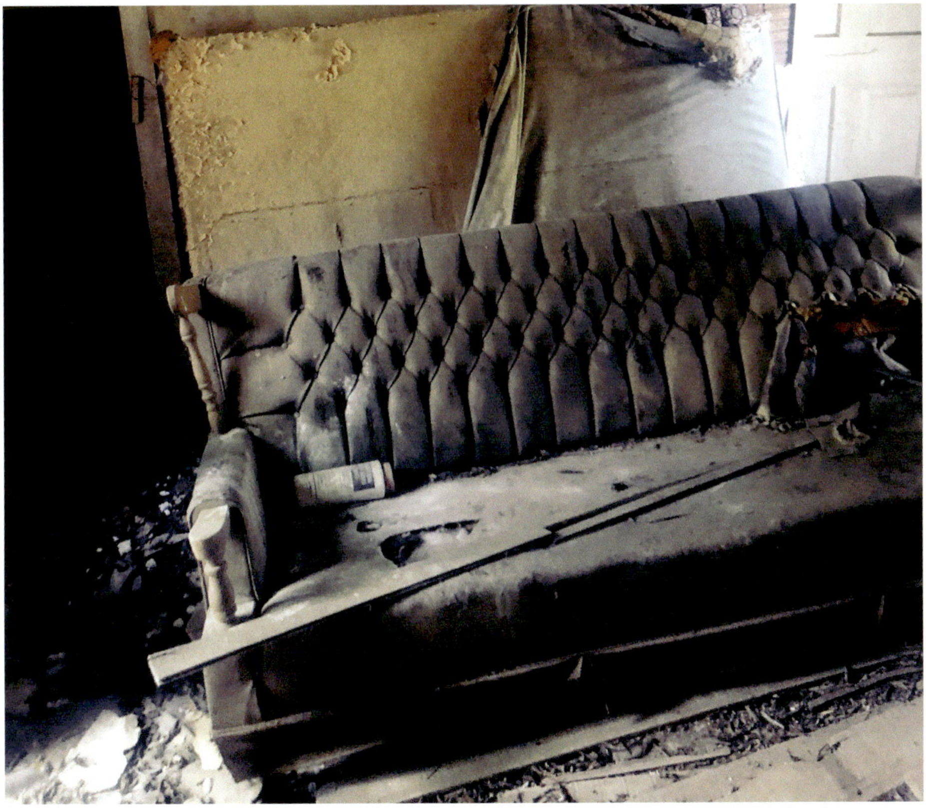

Secrets of the Woods

A weathered house stands connected to a rusting retro blue trailer hiding behind rows of trees that camouflage the lumber of the home. Remnants of life and a family that left everything behind rests off the beaten path of Murfreesboro to be embraced by nature.

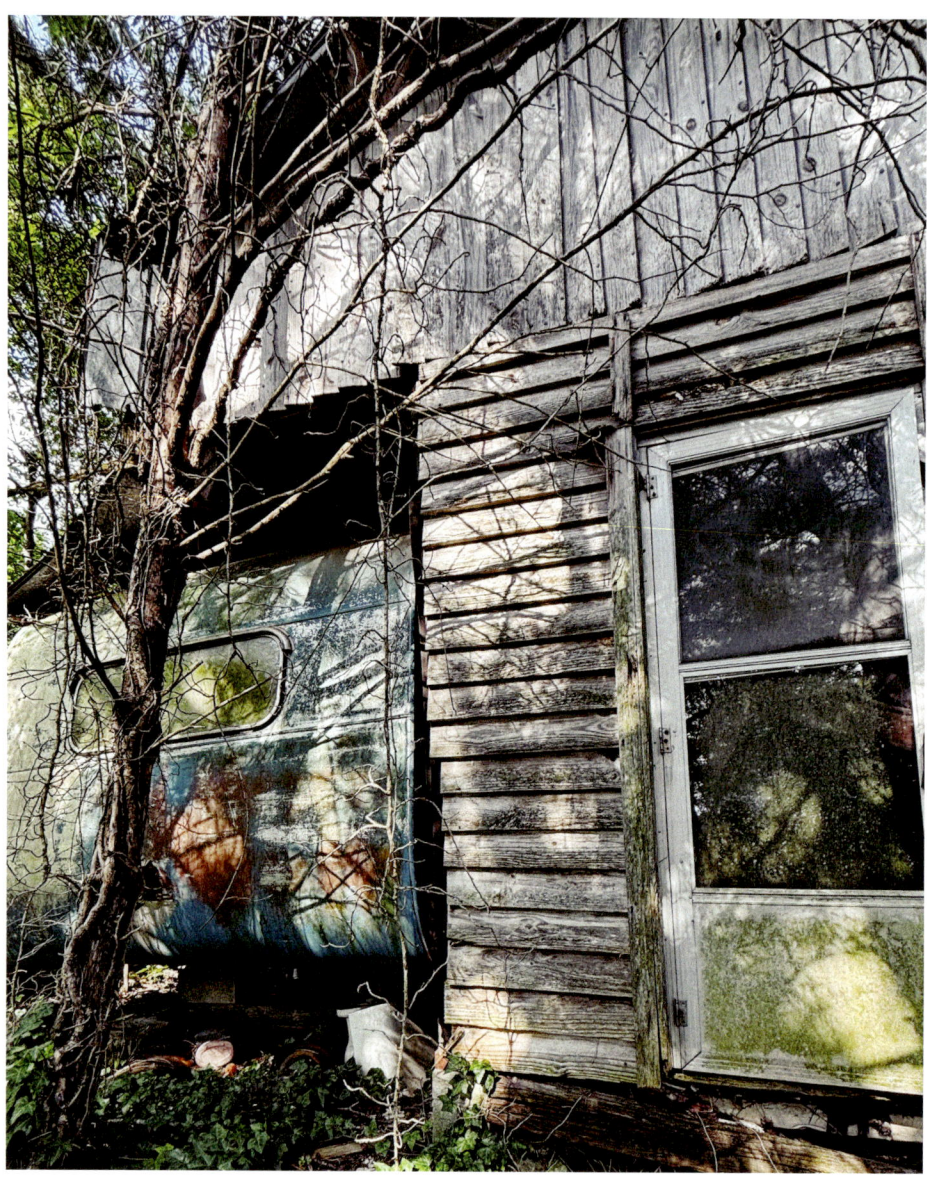

A Chilling Welcome:
Skulls and a Faded Threat Guard a Forgotten Home

This decaying farmhouse whispers tales of neglect. A rusted "attack dog" sign mocks emptiness. Weathered cow skulls stand guard on the fence. Nature reclaims the land, beauty blooming amid decay.

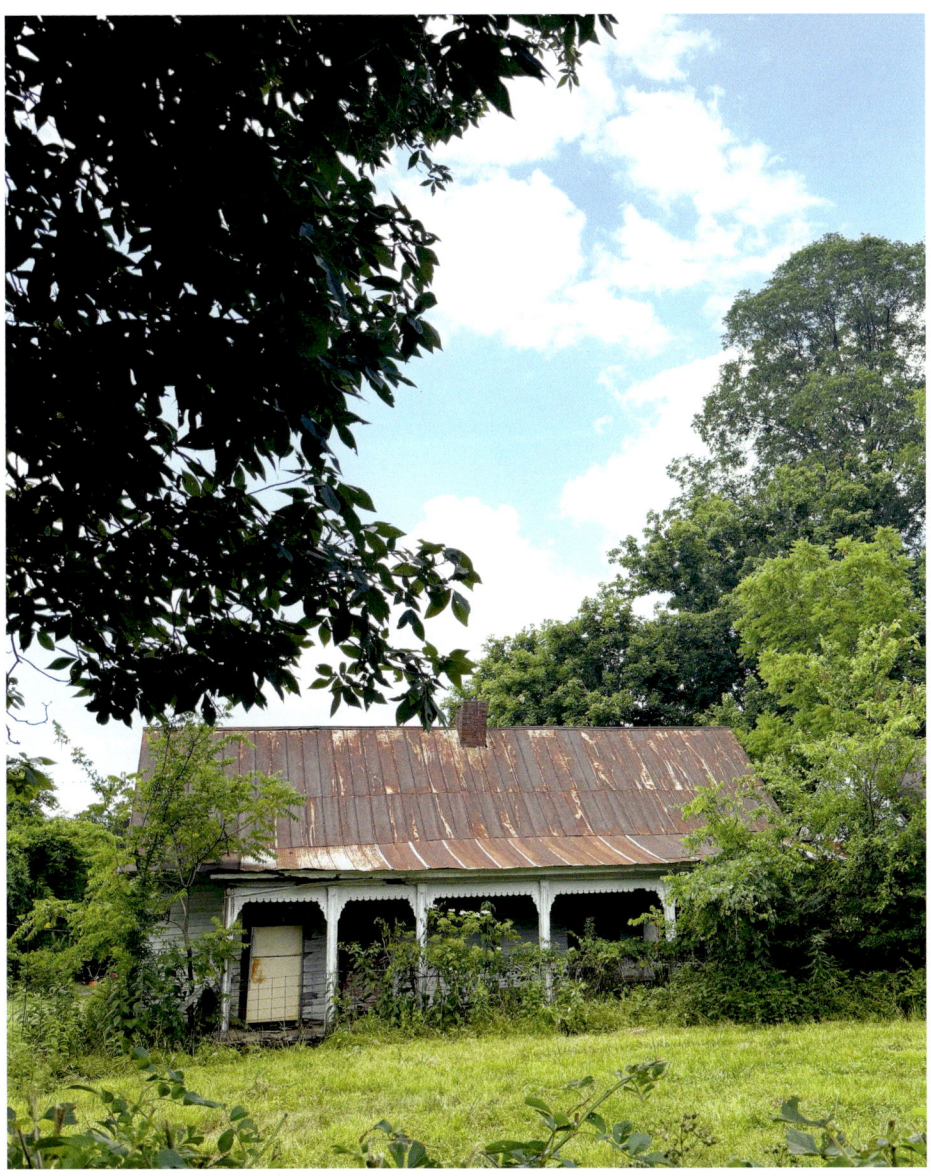

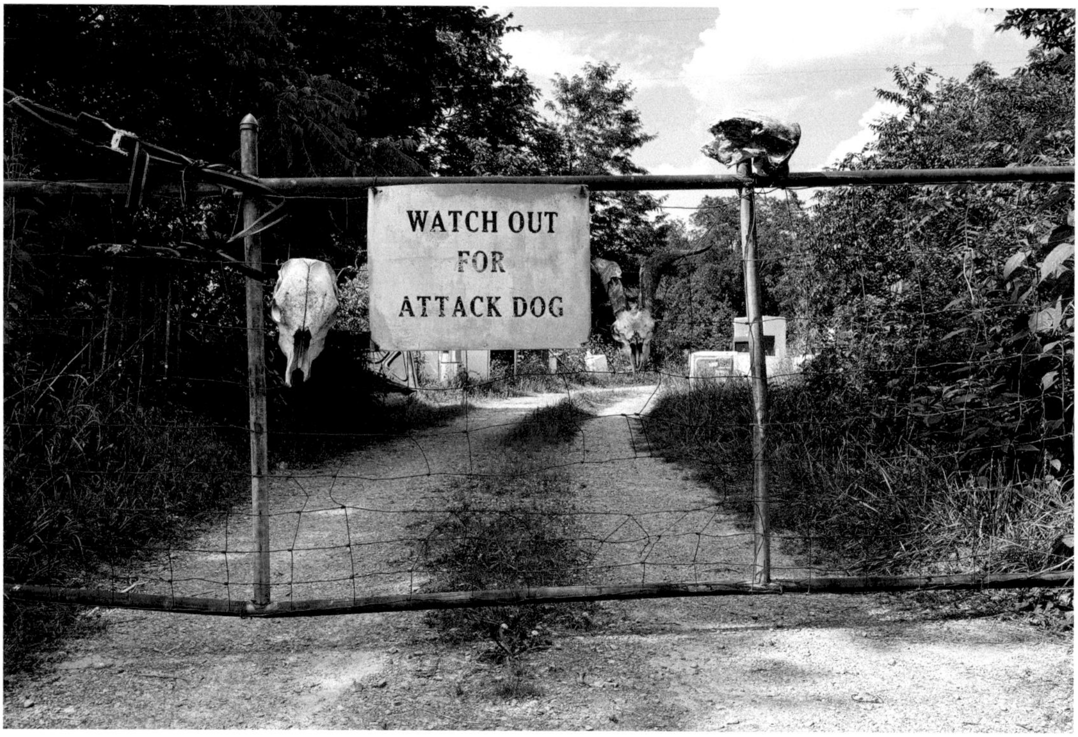
WATCH OUT
FOR
ATTACK DOG

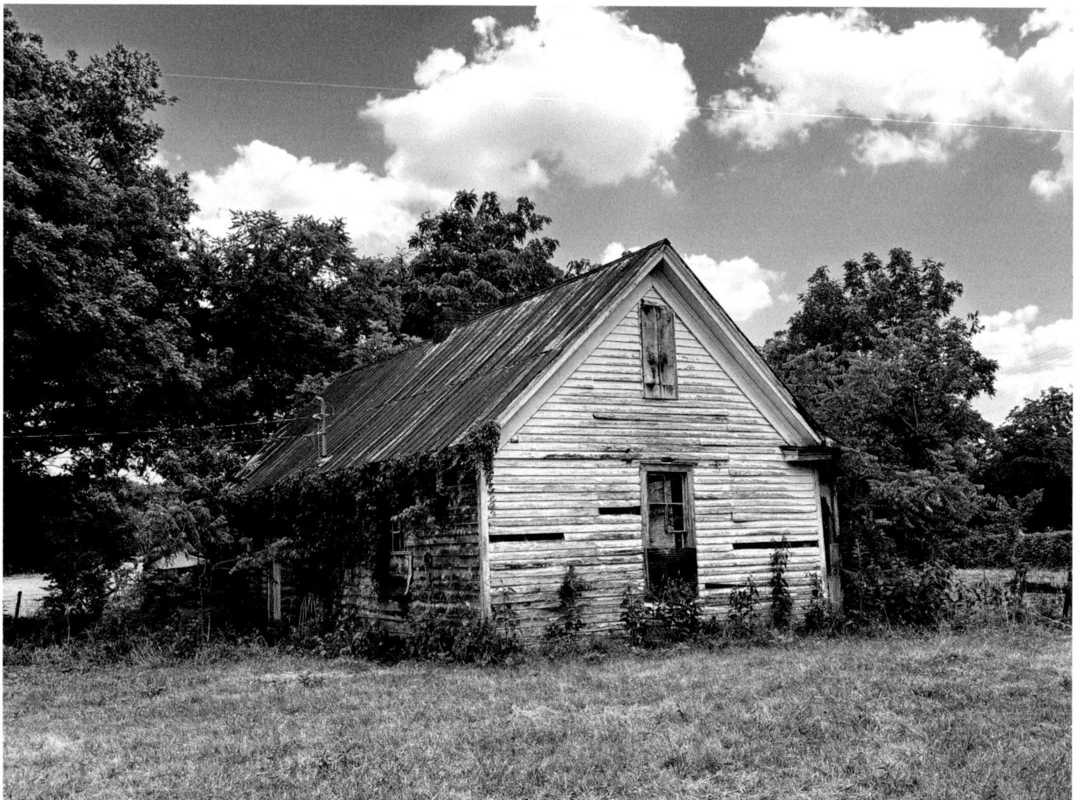

Where Education Bloomed, or Families Grew?
Time Blurs the Lines in this Forgotten Structure

Forgotten on the farm, a weathered building holds its purpose close. Schoolhouse or family home, only whispers remain. Sunlight paints emptiness within, while a faded blackboard remembers lessons learned. Nature reclaims, weaving beauty from a forgotten era's secrets.

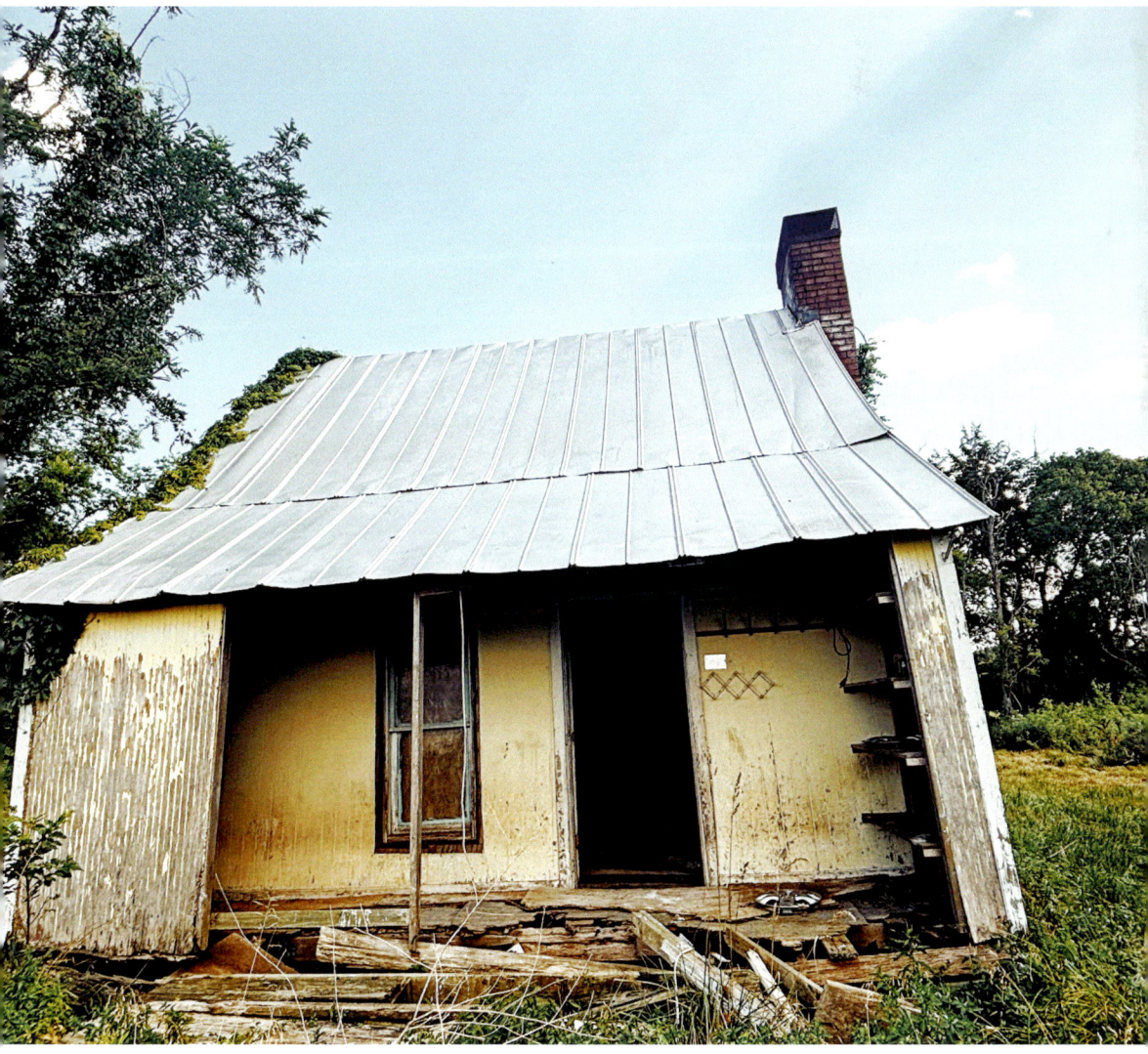

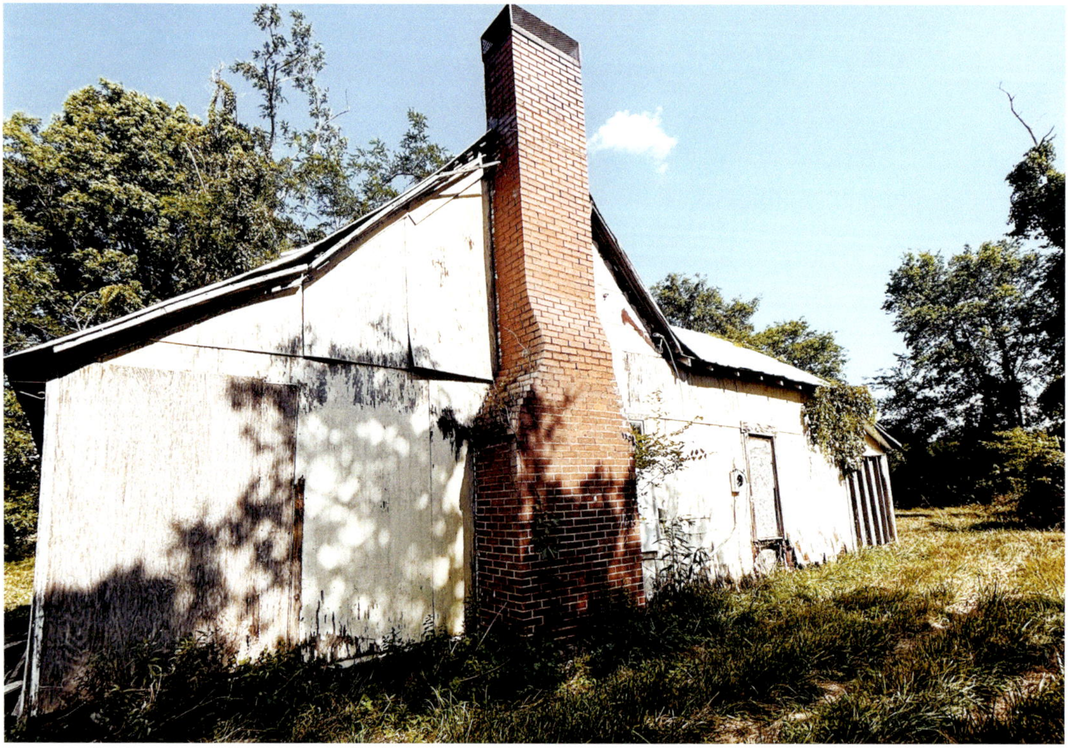

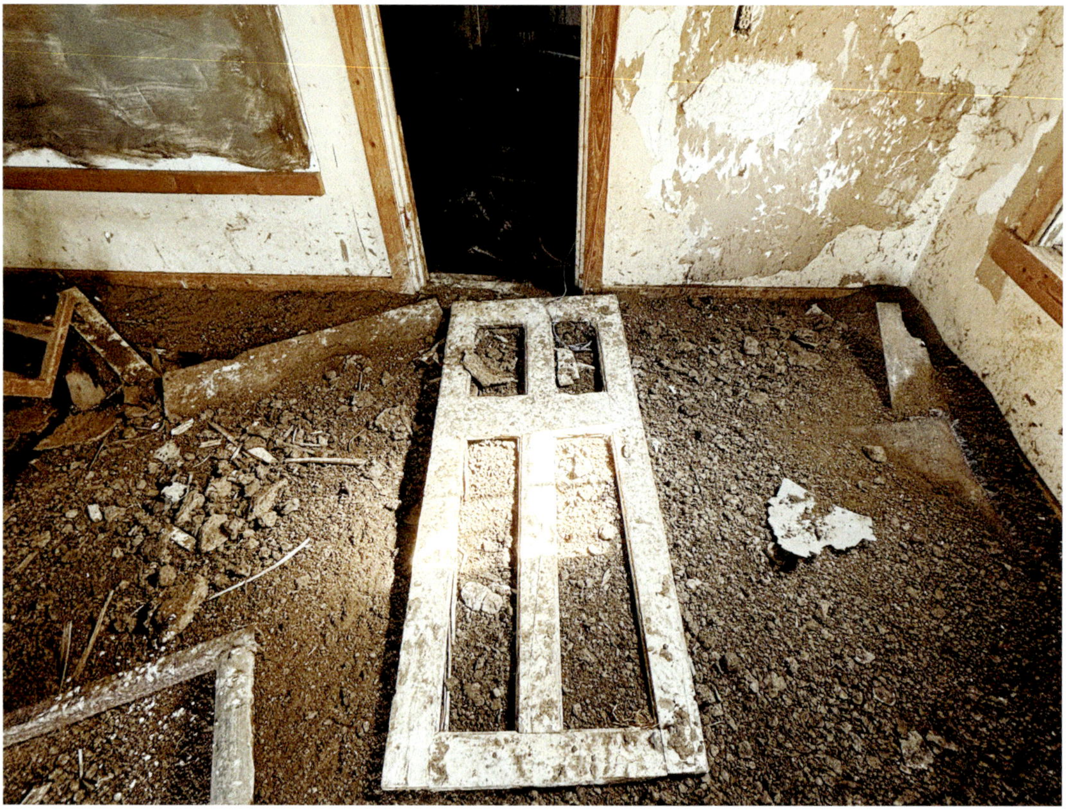

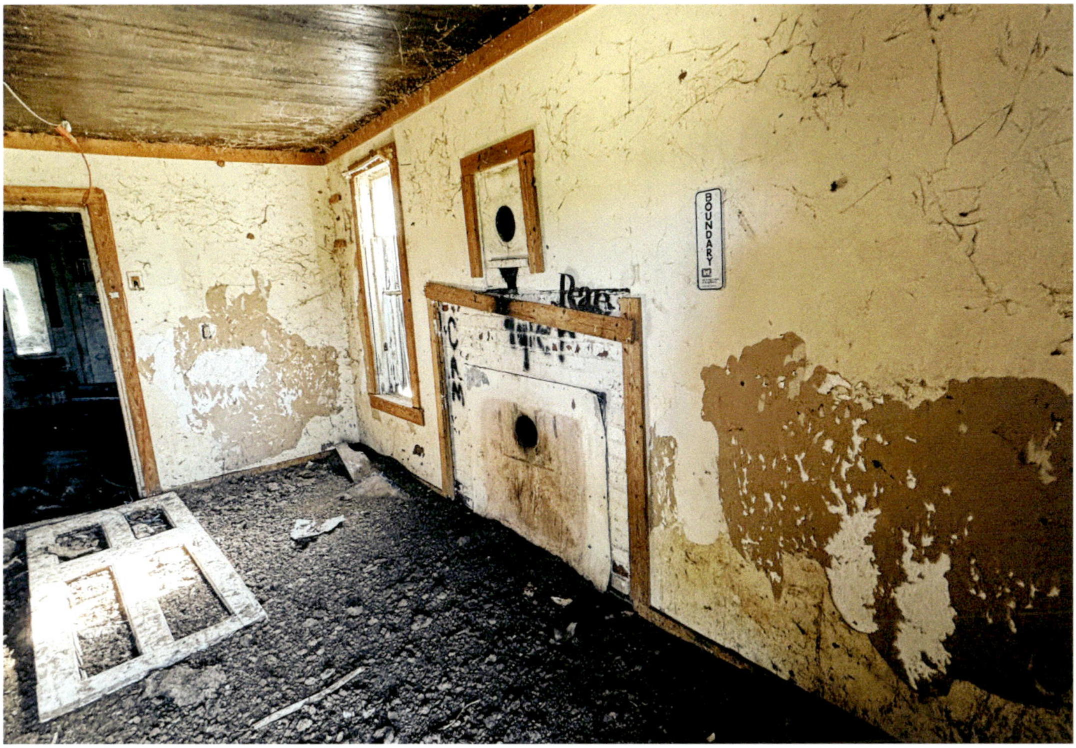
BOUNDARY

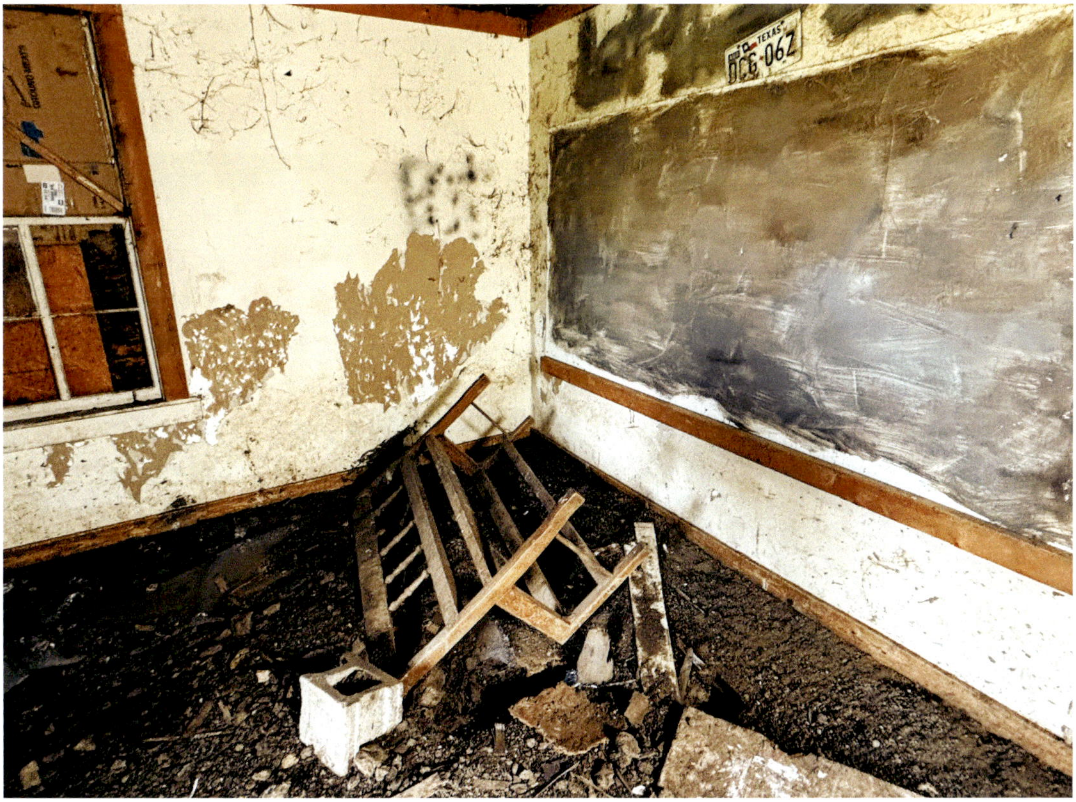
TEXAS
DCG-06Z

Nature's Embrace: A Forgotten Home Surrenders to the Vines

Vines win the battle, engulfing a forgotten farmhouse. Siding whispers, the chimney stands tall—silent testament to a home reclaimed by nature.

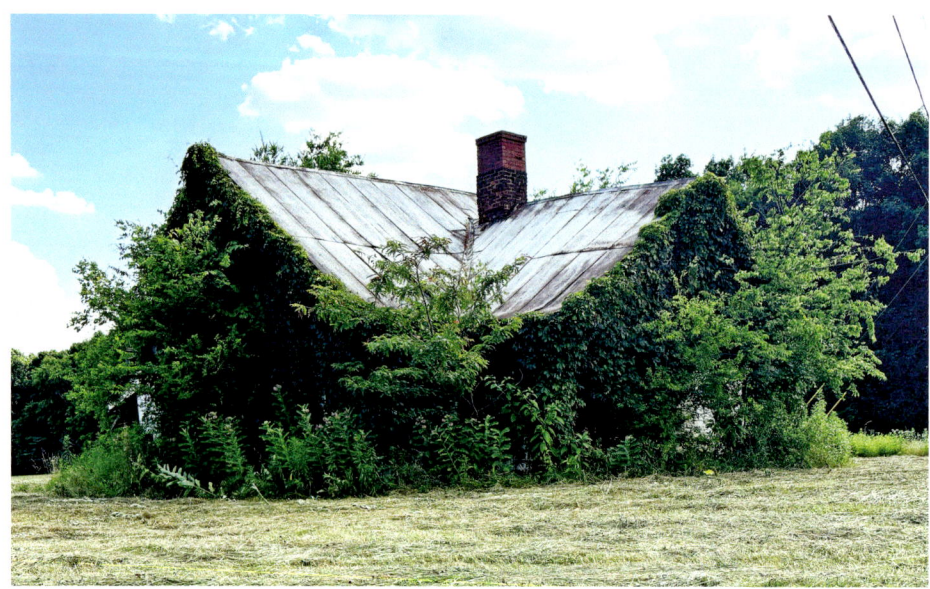

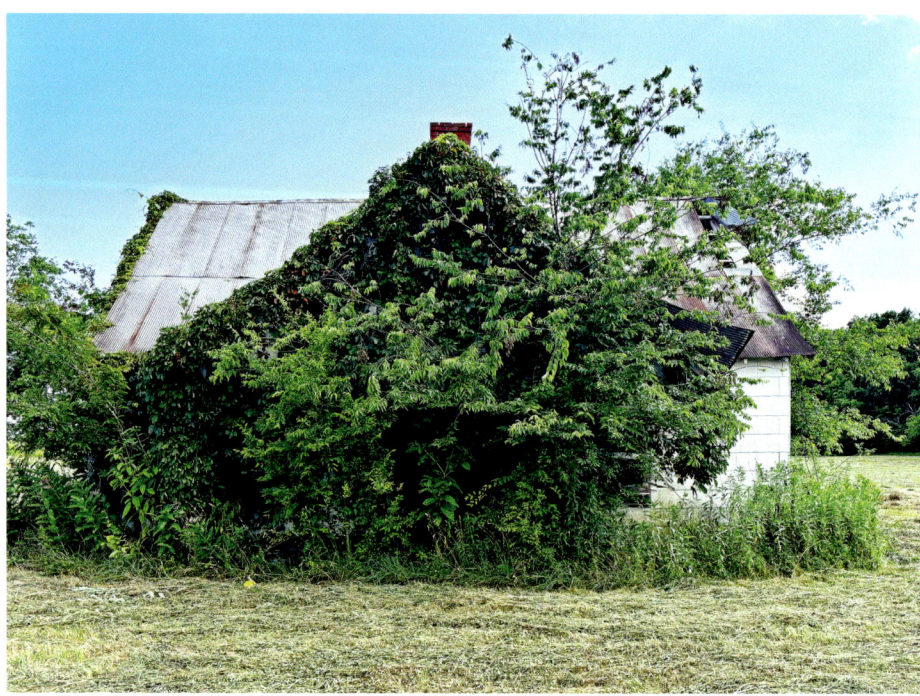

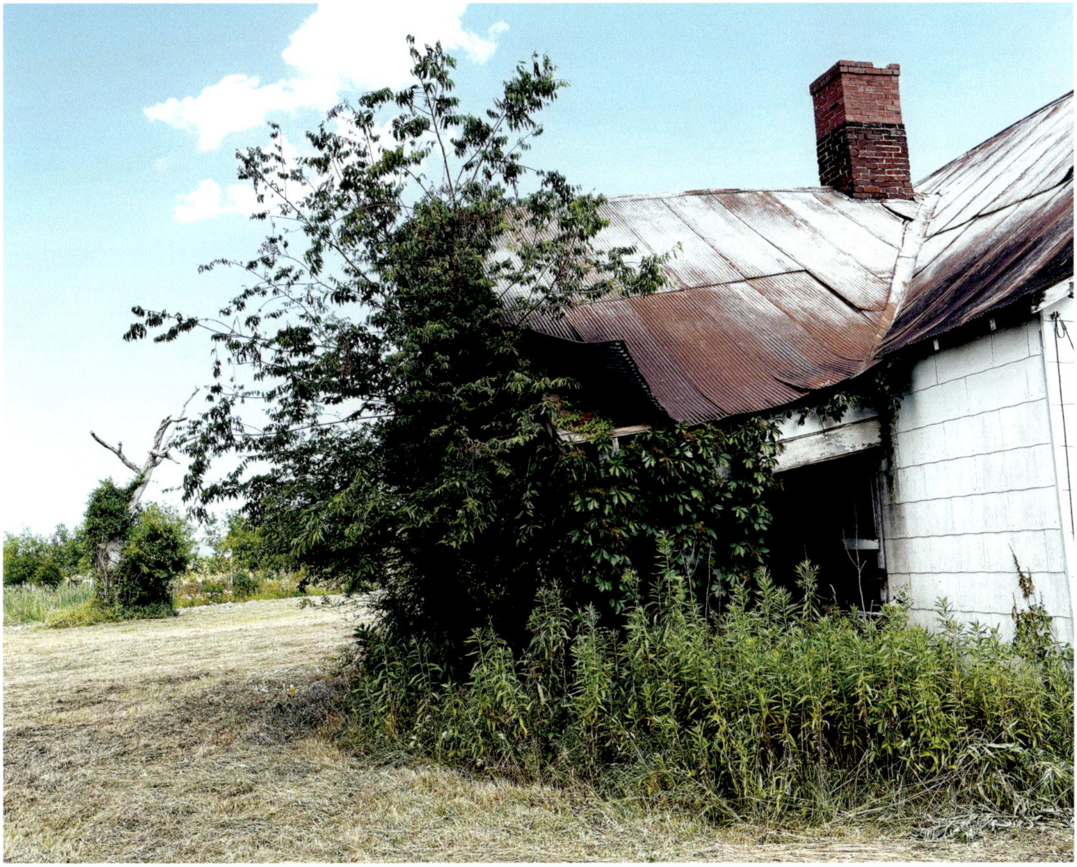

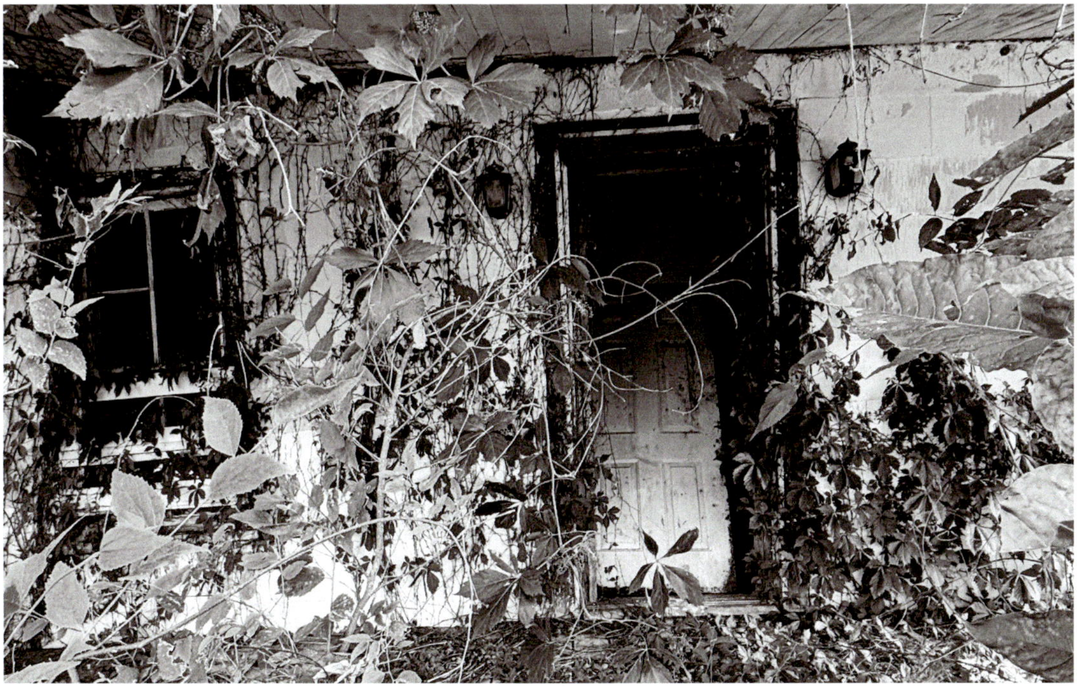

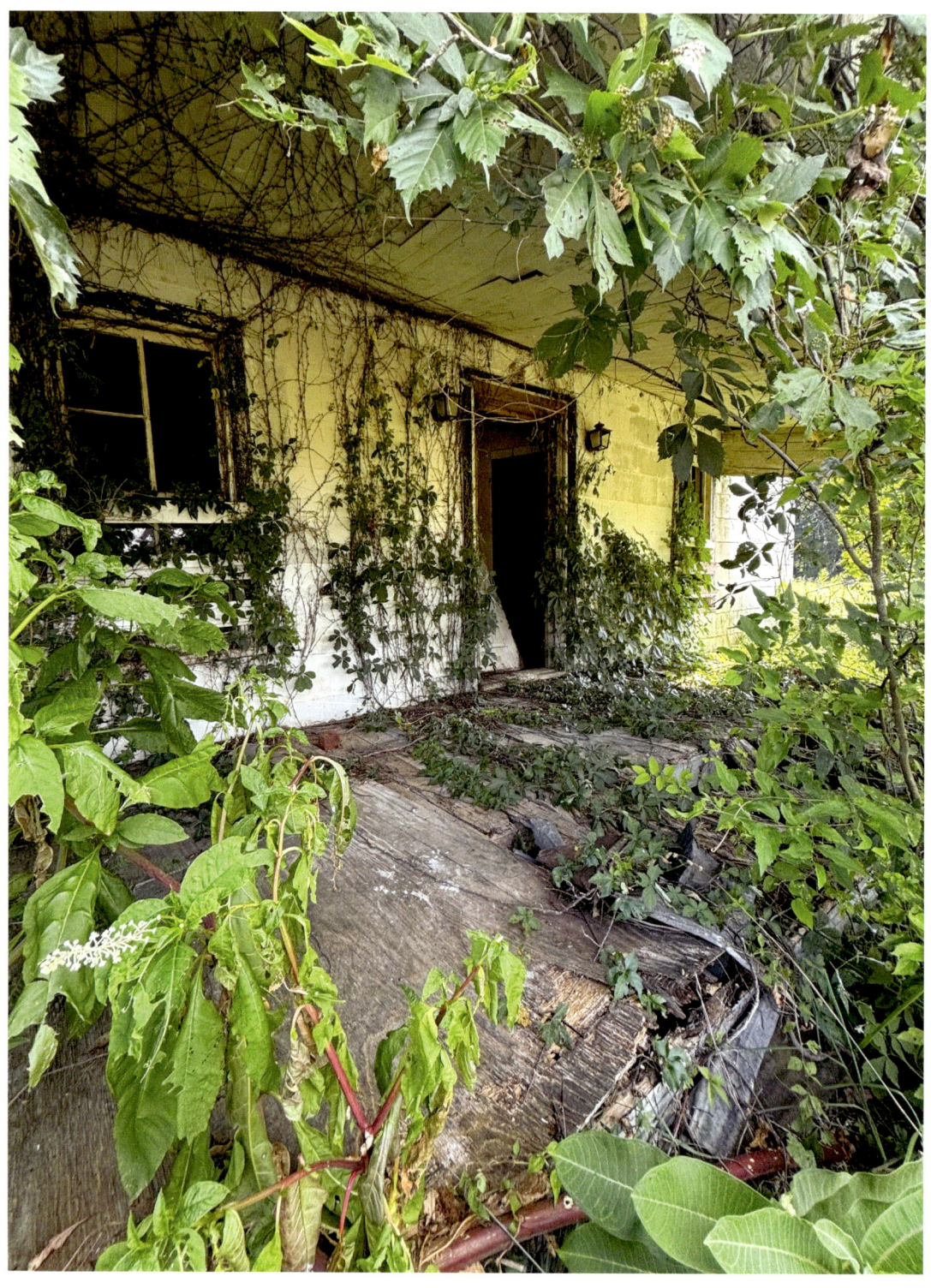

Echoes of Stillness

Sunlight paints a forgotten farmhouse. A twisted tree guards the porch, where rocking chairs whisper of lives long gone. Nature reigns, memories linger.

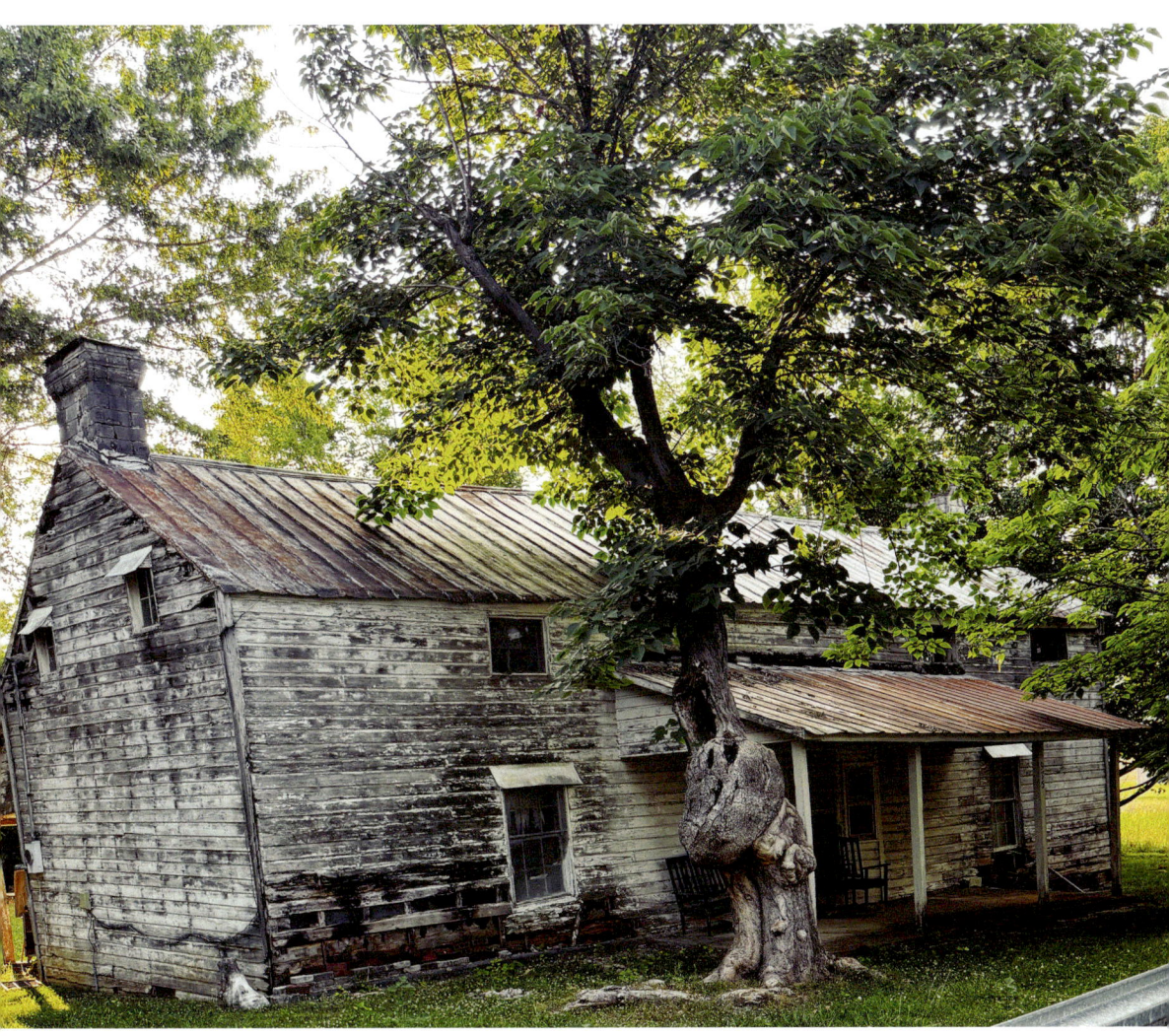

Nature Reclaims the Harvest

Sun paints a lonely farmhouse. Rusted milk jug and rack stand silent—a forgotten symbol of a farm's bounty. Shadows whisper inside, hinting at untold stories.

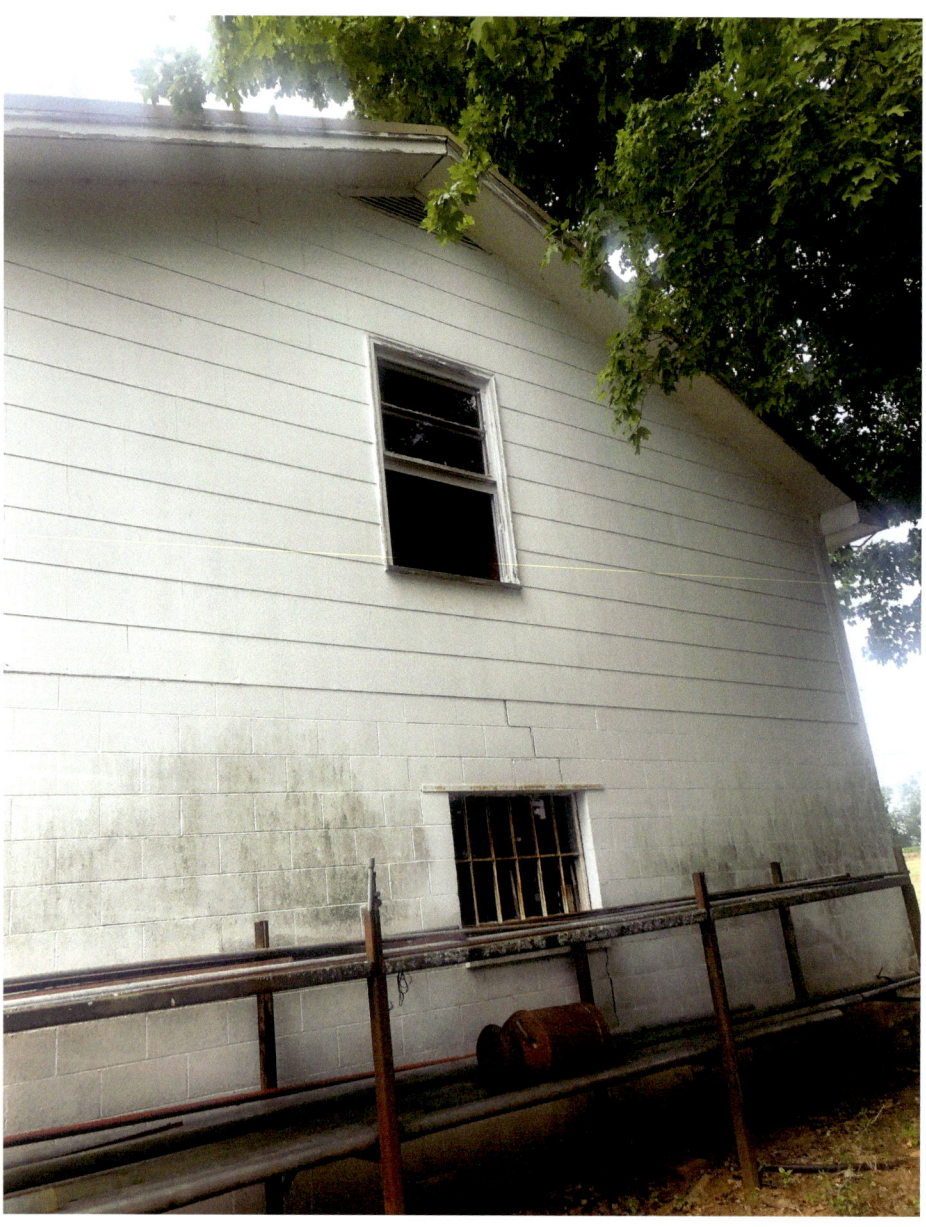

Echoes of Grandeur: A Fallen Mansion Whispers of Secrets

A once-grand mansion, now abandoned due to the family's misfortune, rots behind overgrown foliage and a broken white picket fence; a metaphor to a family's American dream crumbling due to financial hardship, divorce, and death. Lavish interior hints at former life, while vandalism marks the passage of time.

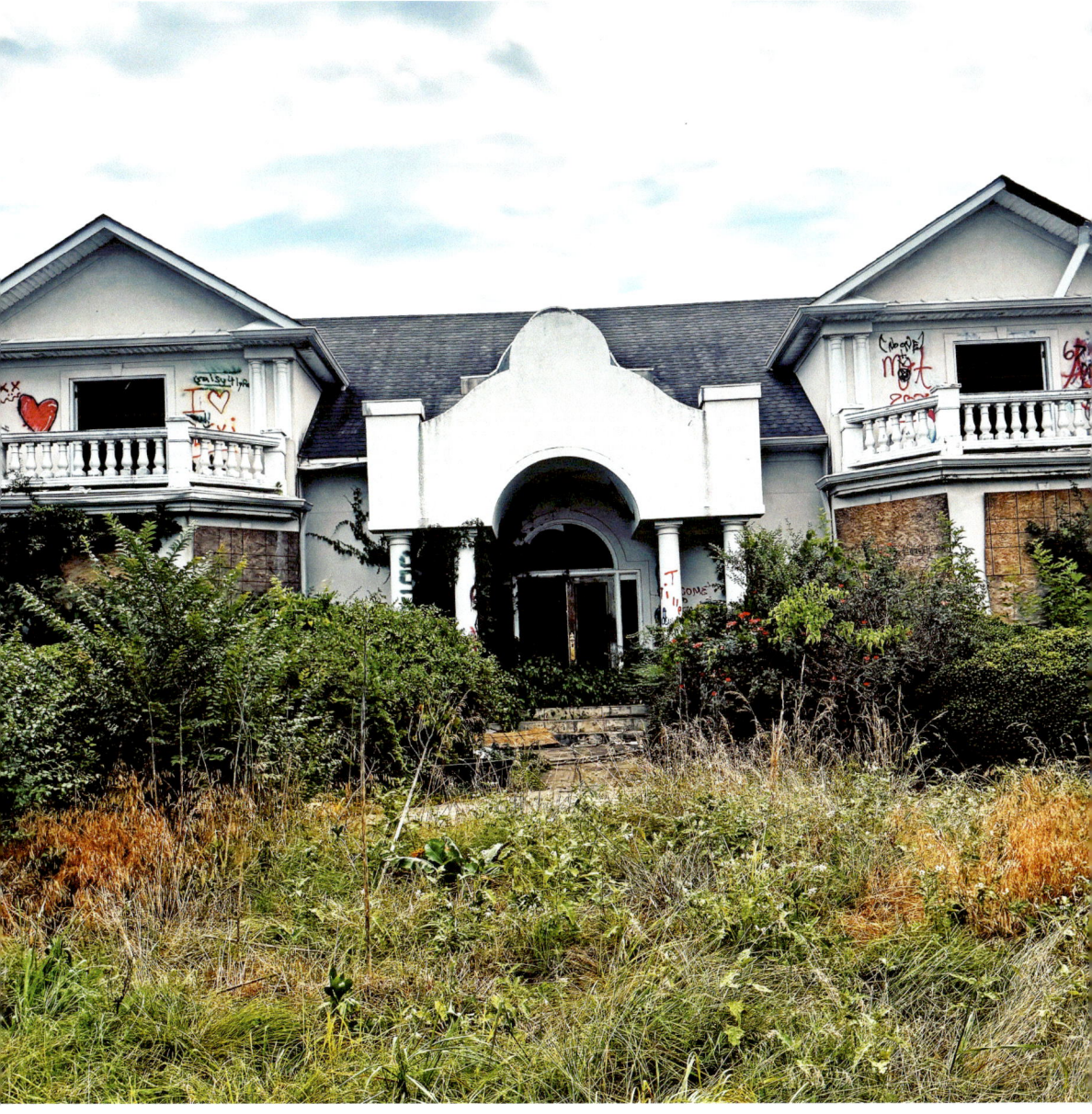

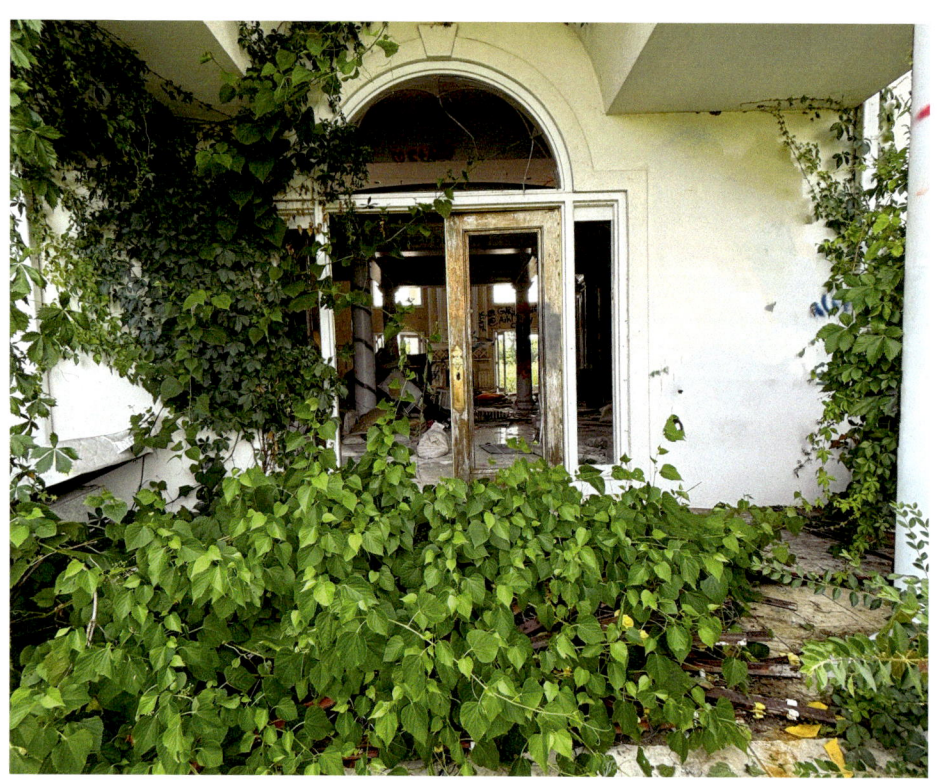

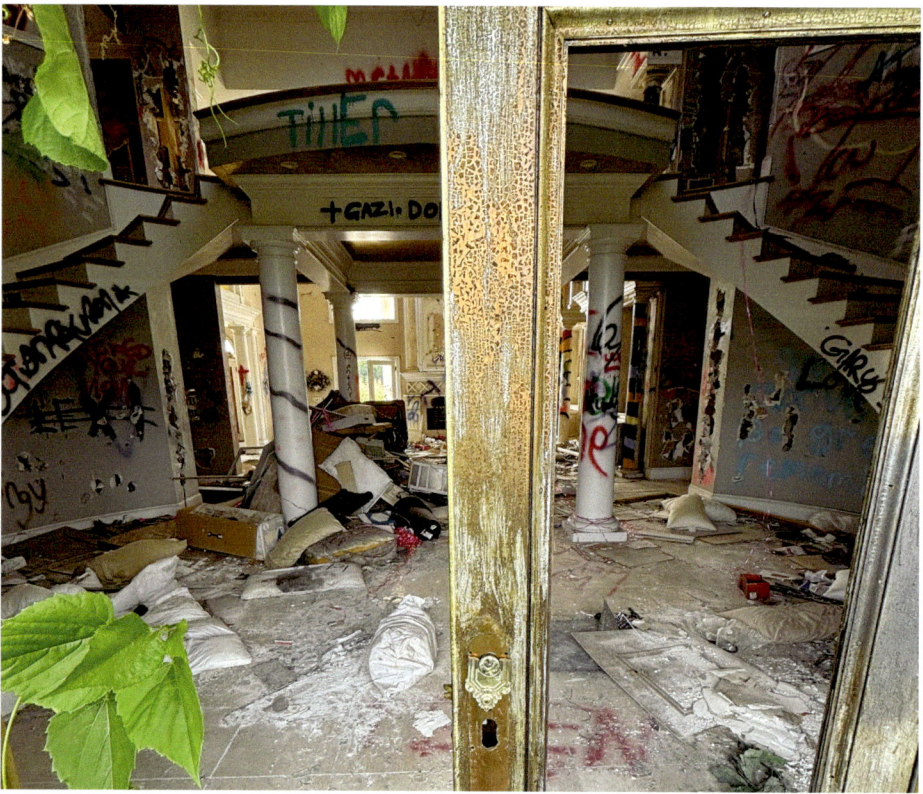

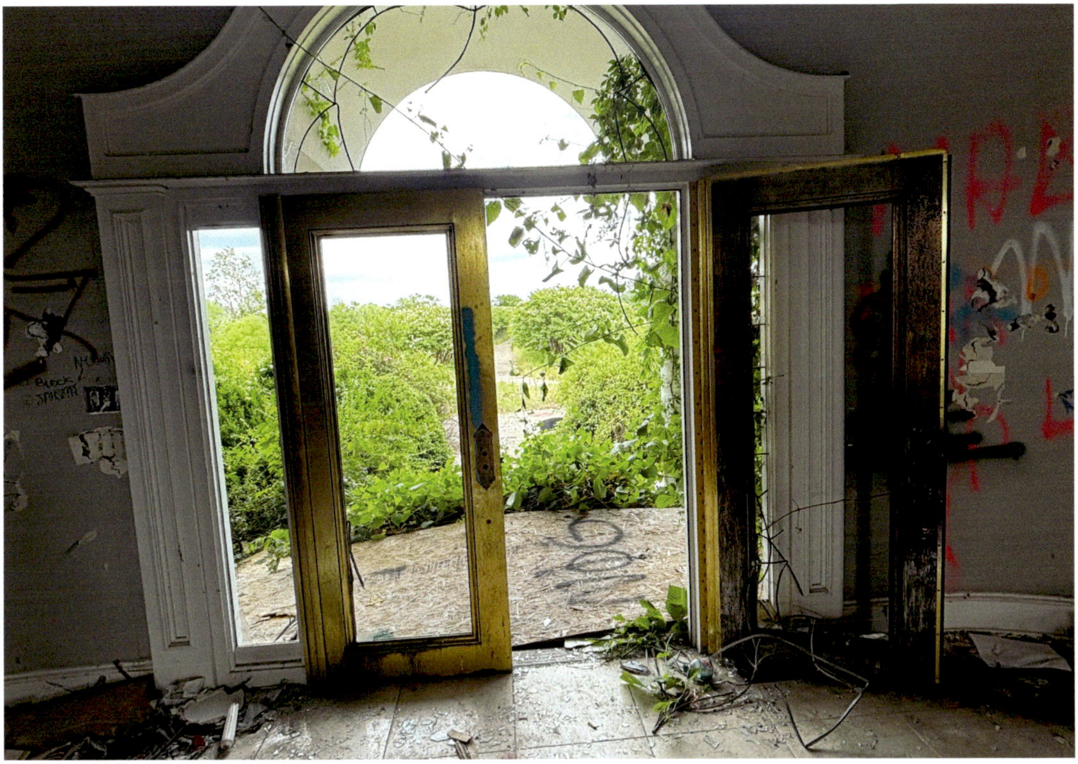

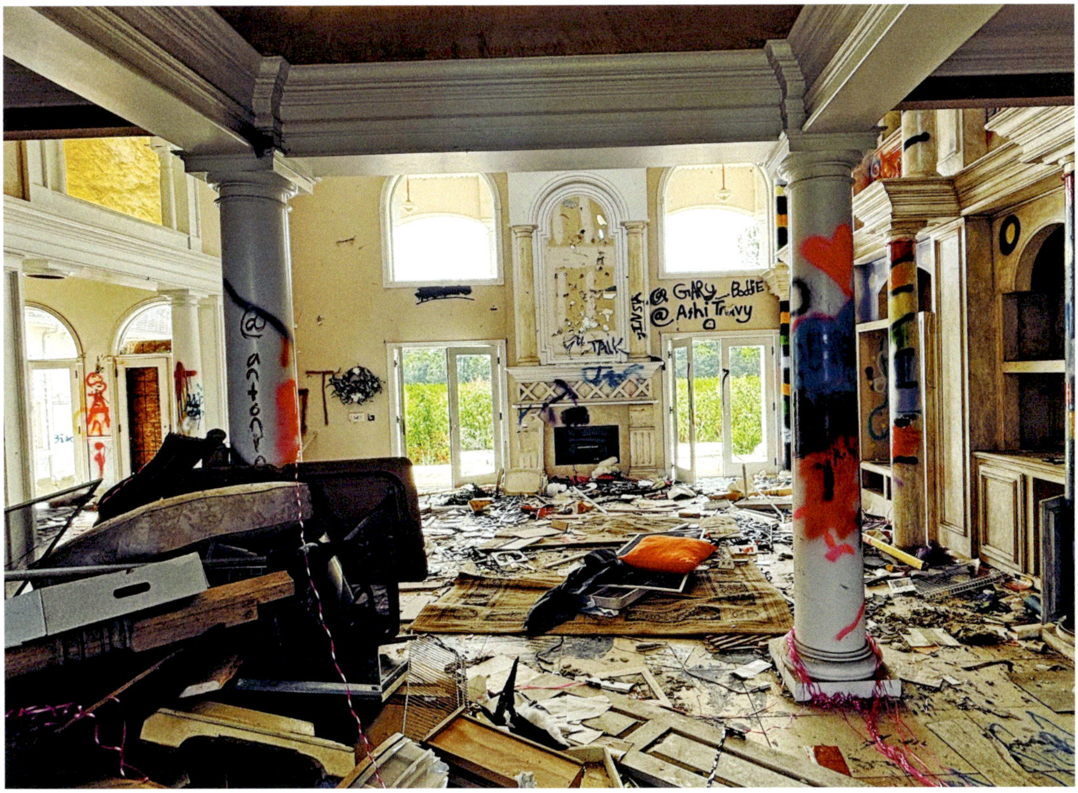
INSTA
@ GARY_BoddE
@ Ashi Truvy
TALK

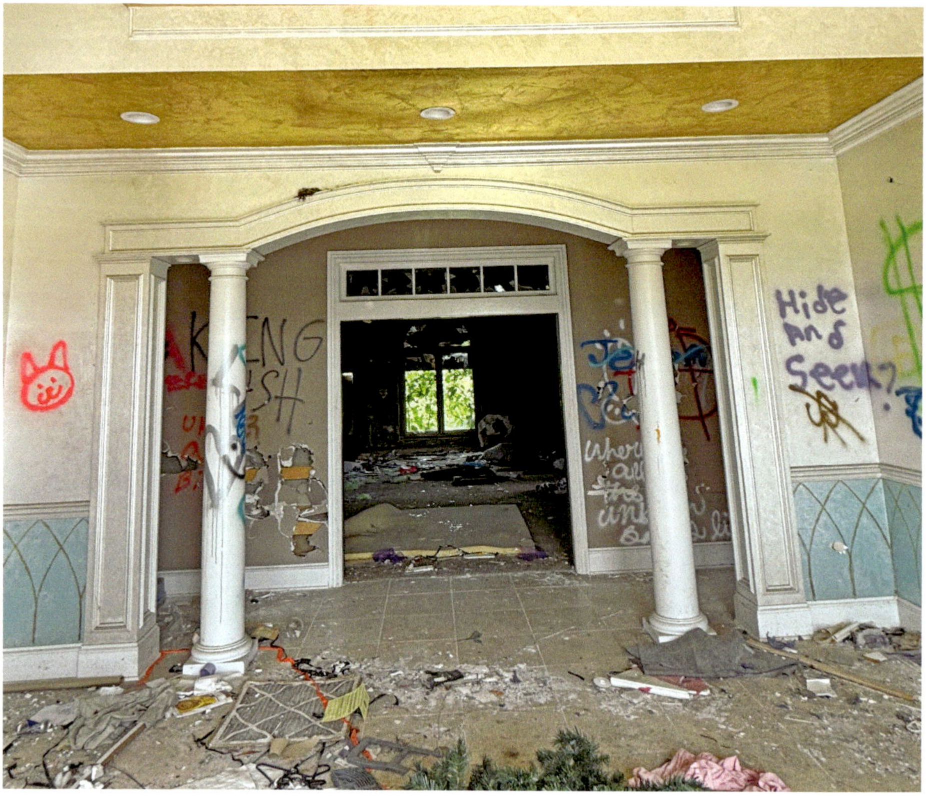

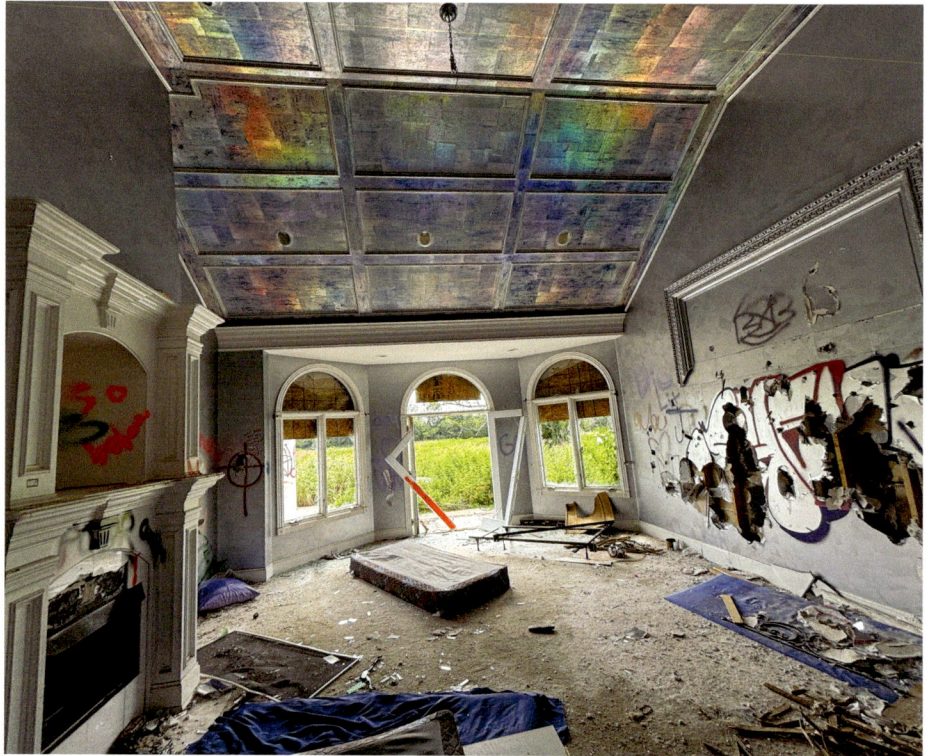

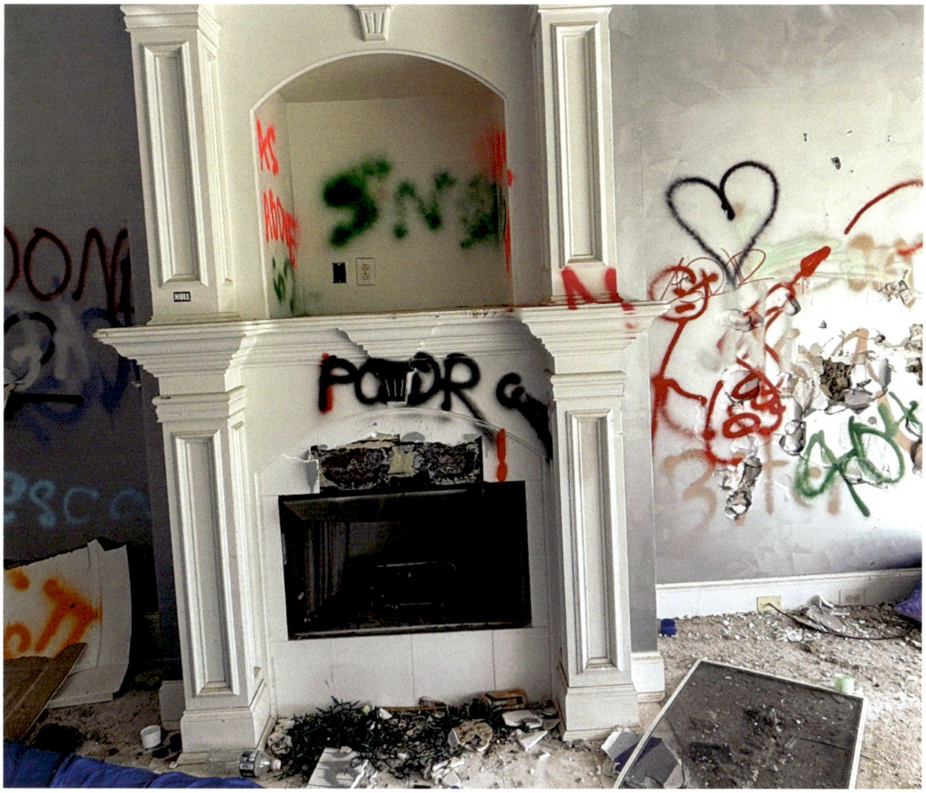

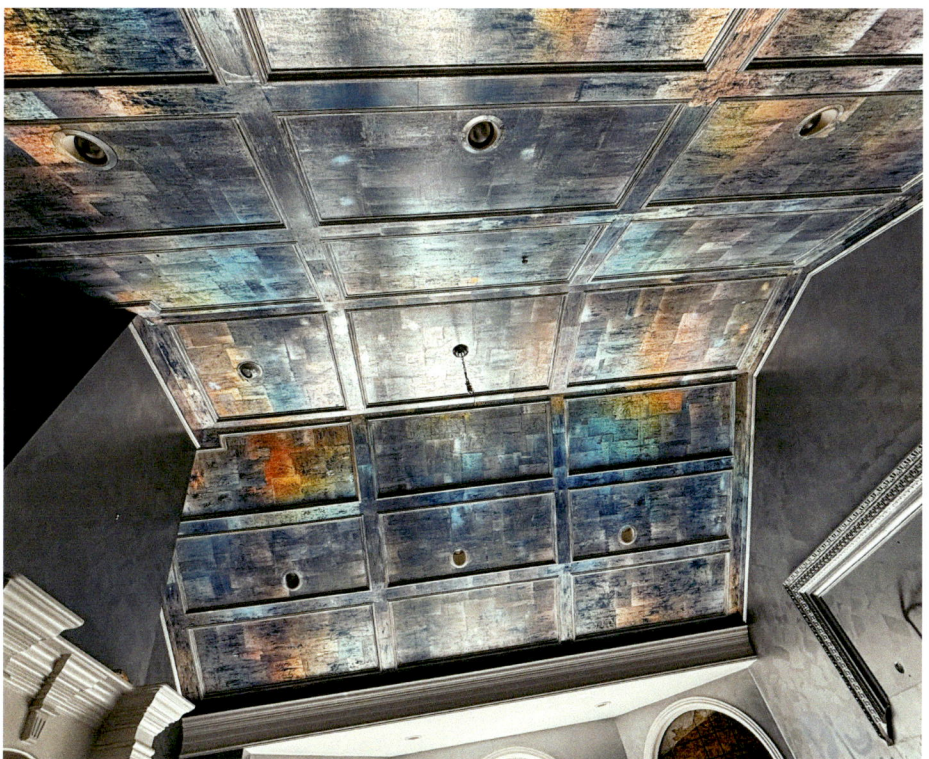

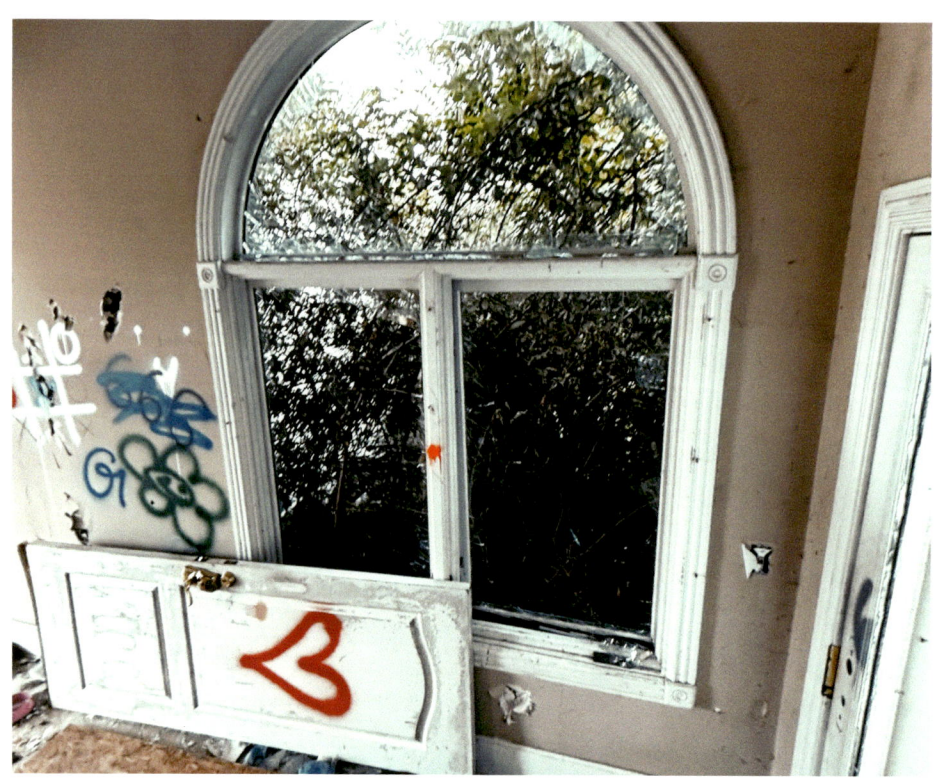

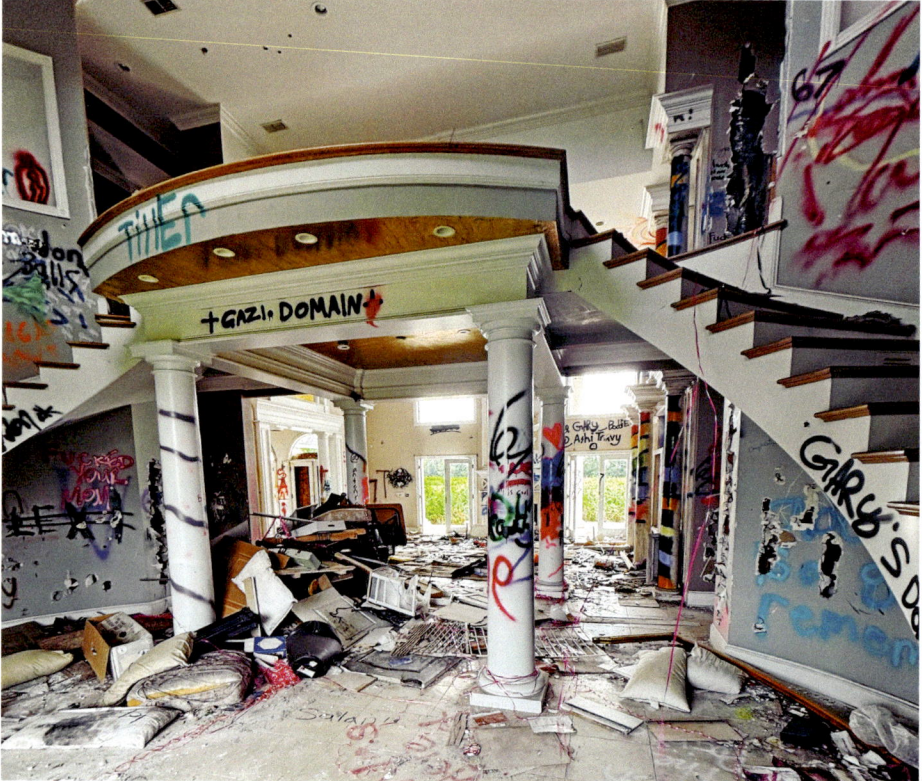

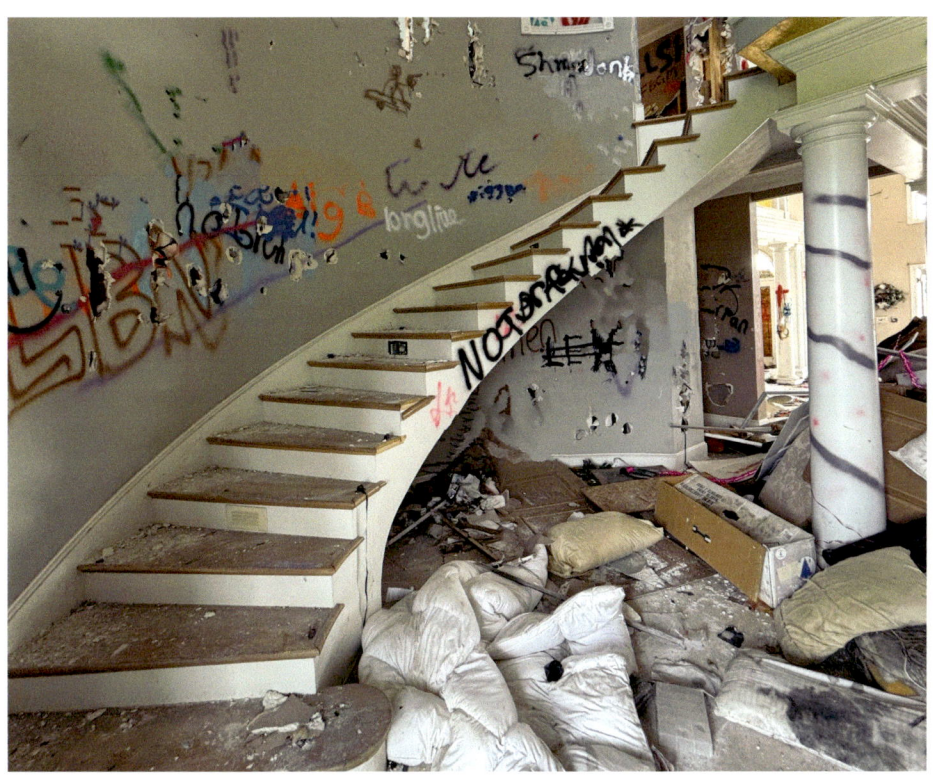

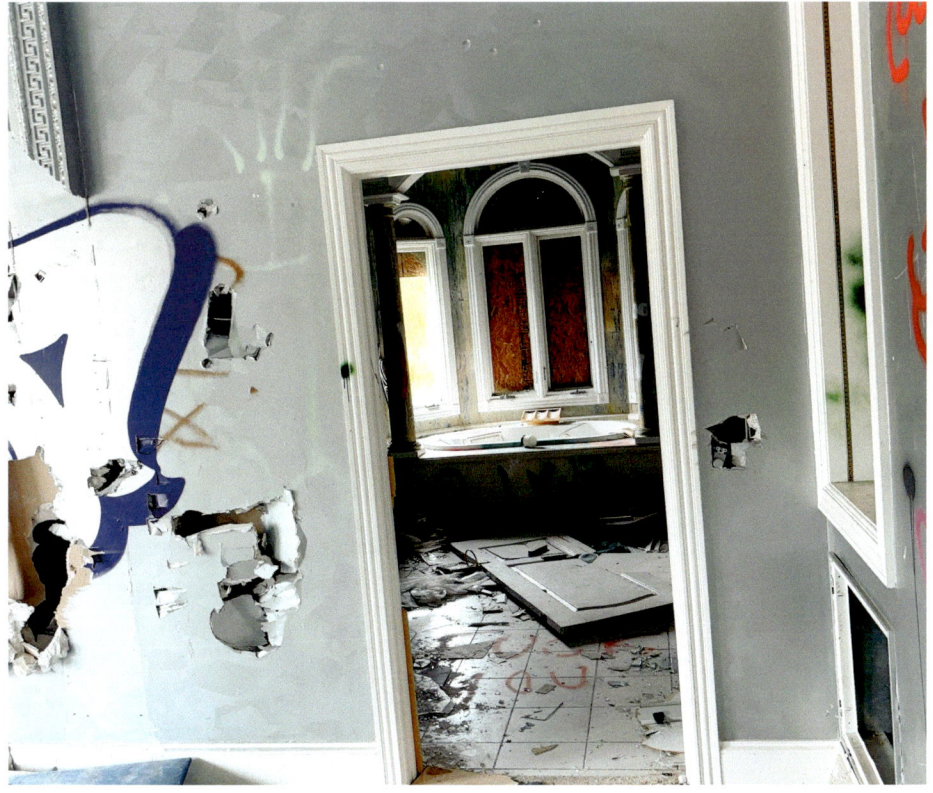

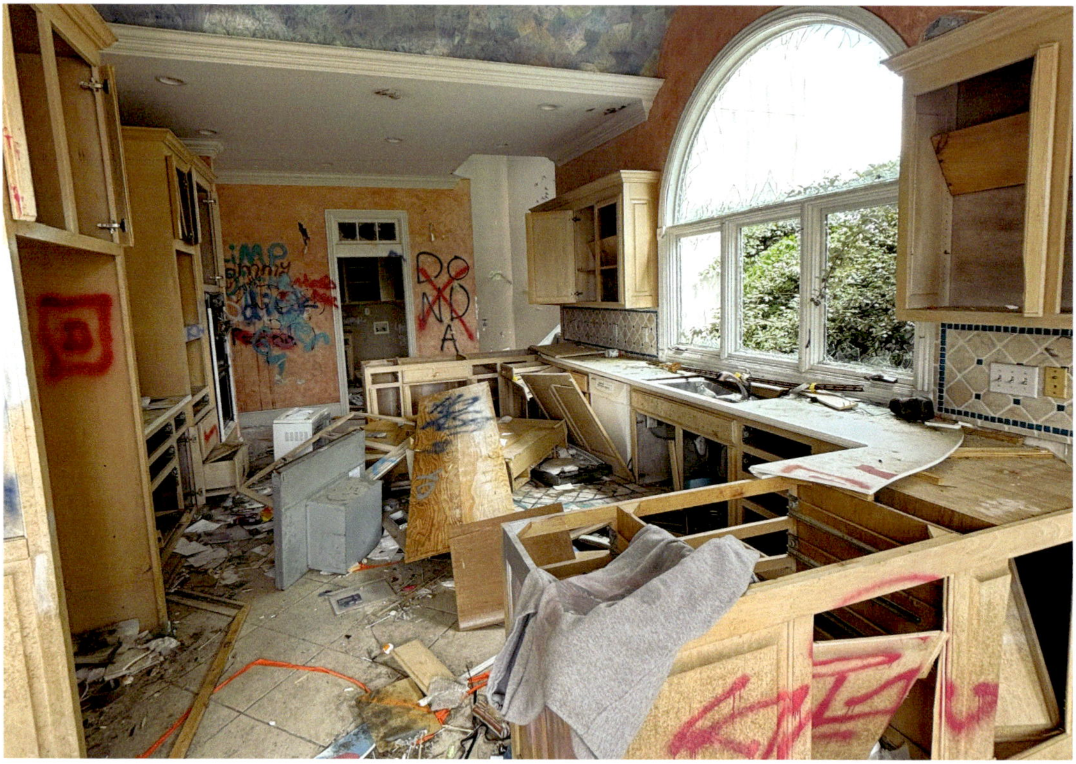

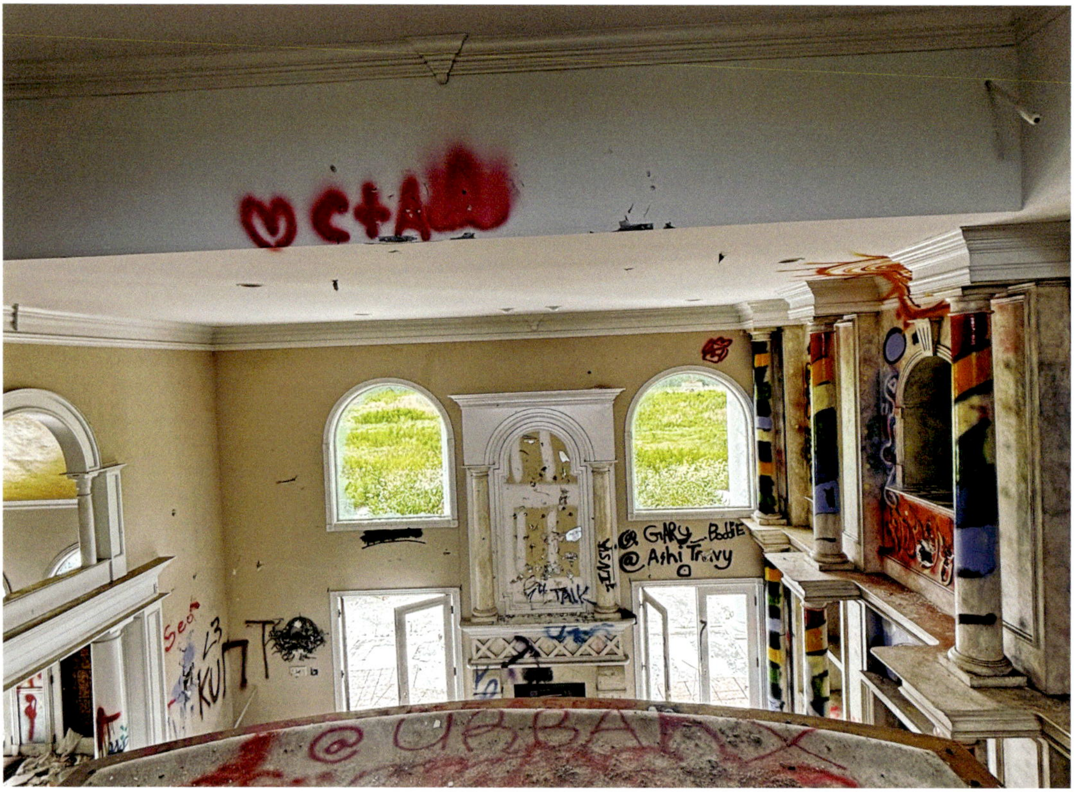

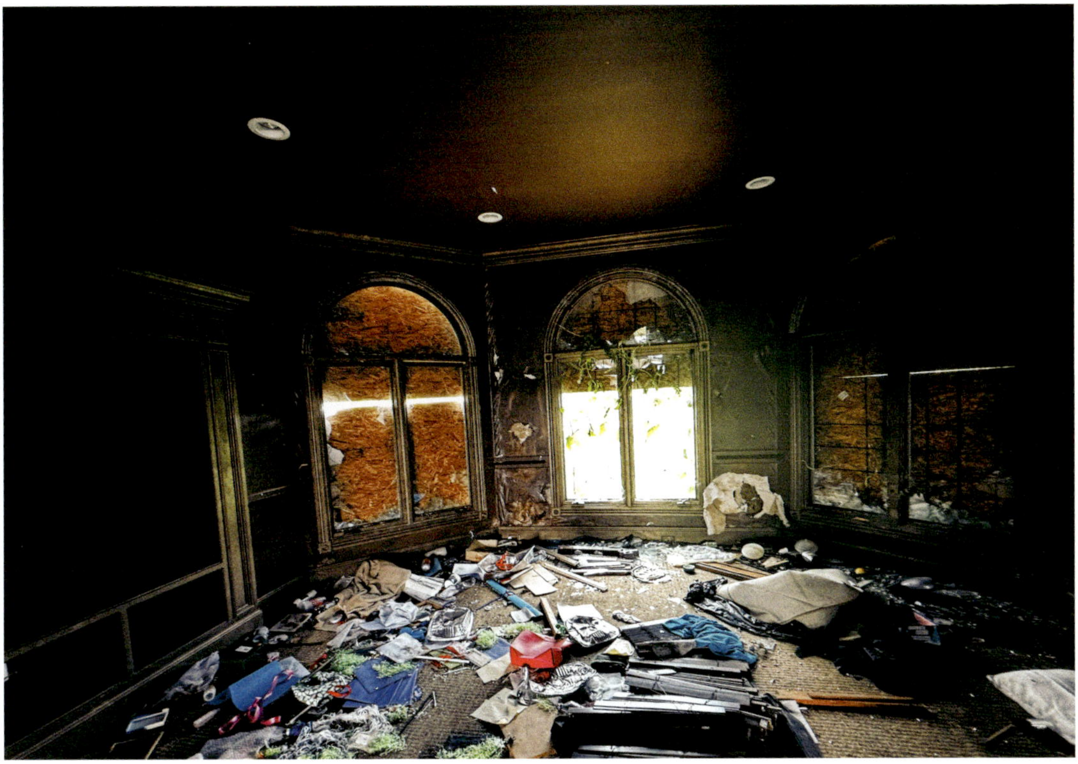

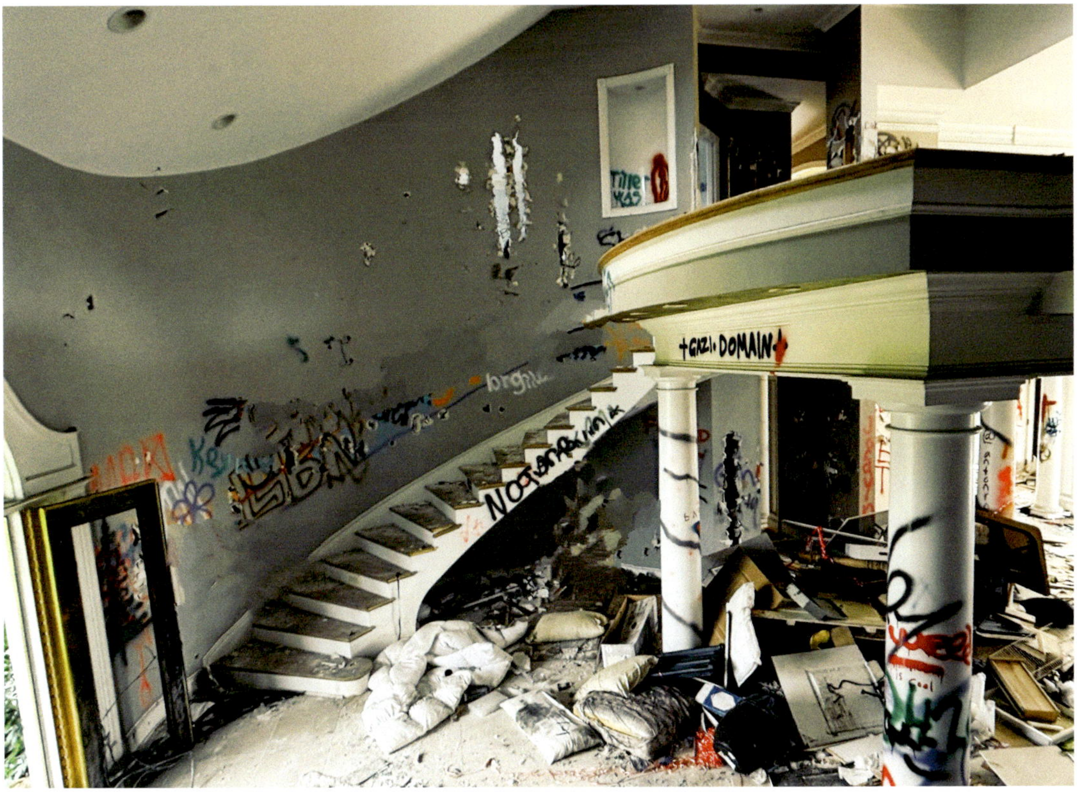
GAZI·DOMAIN

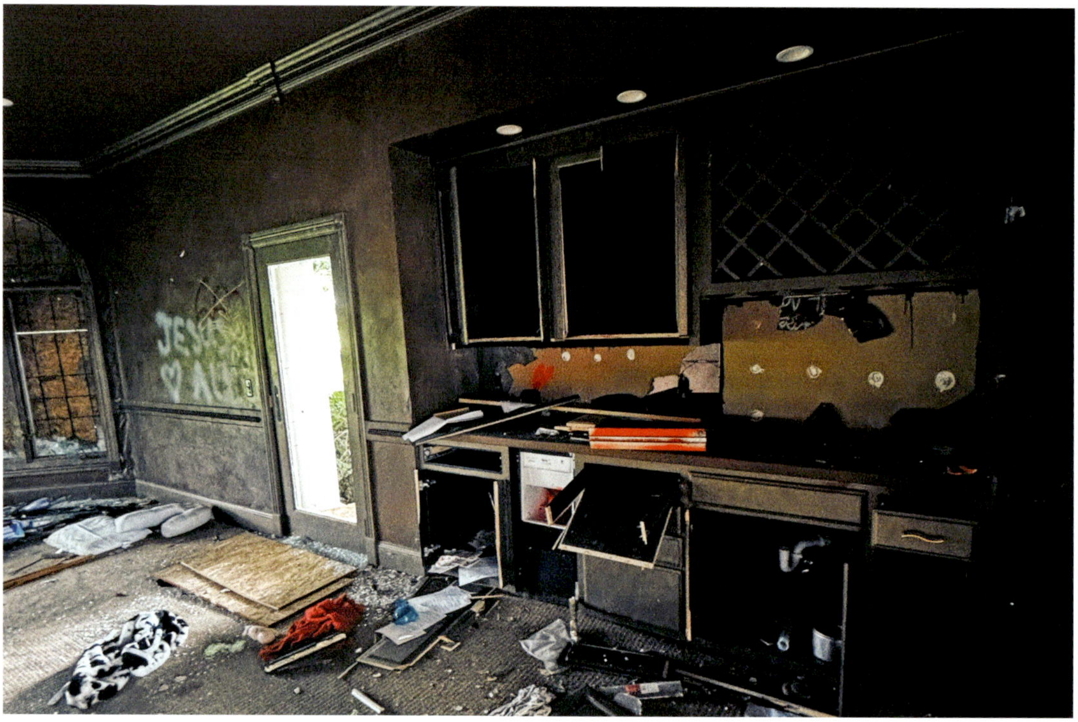

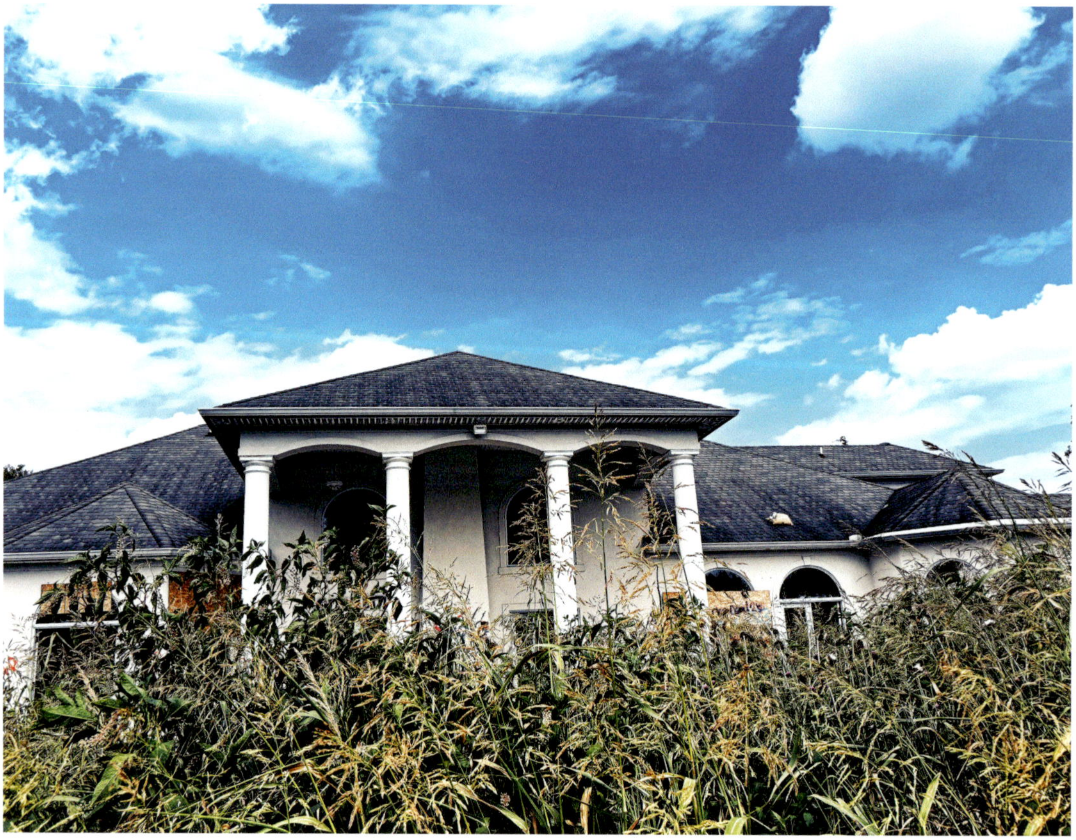

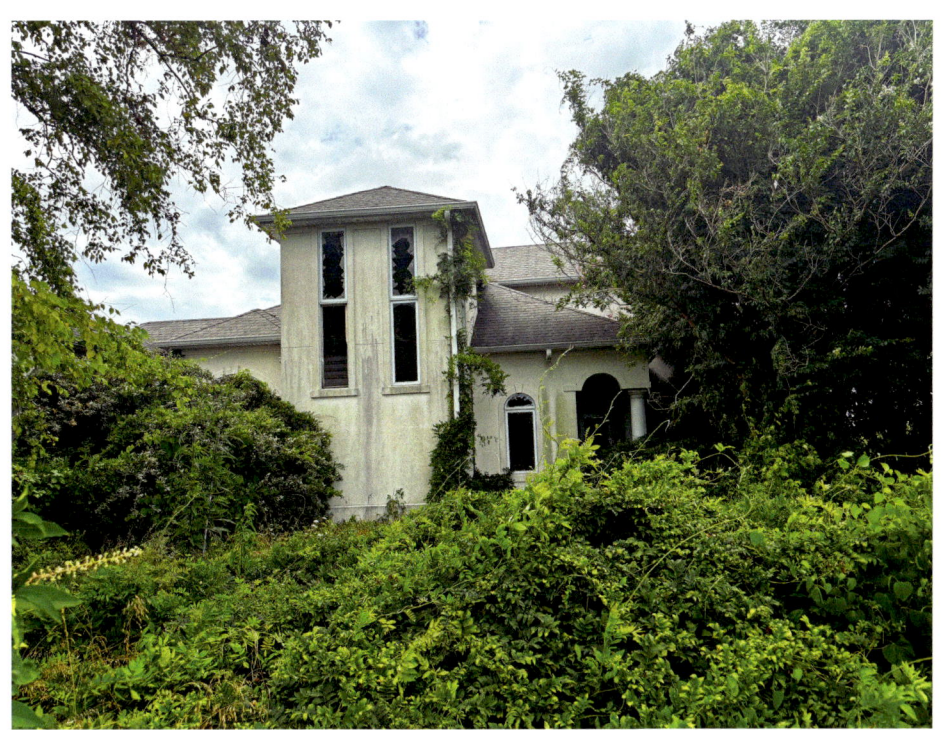

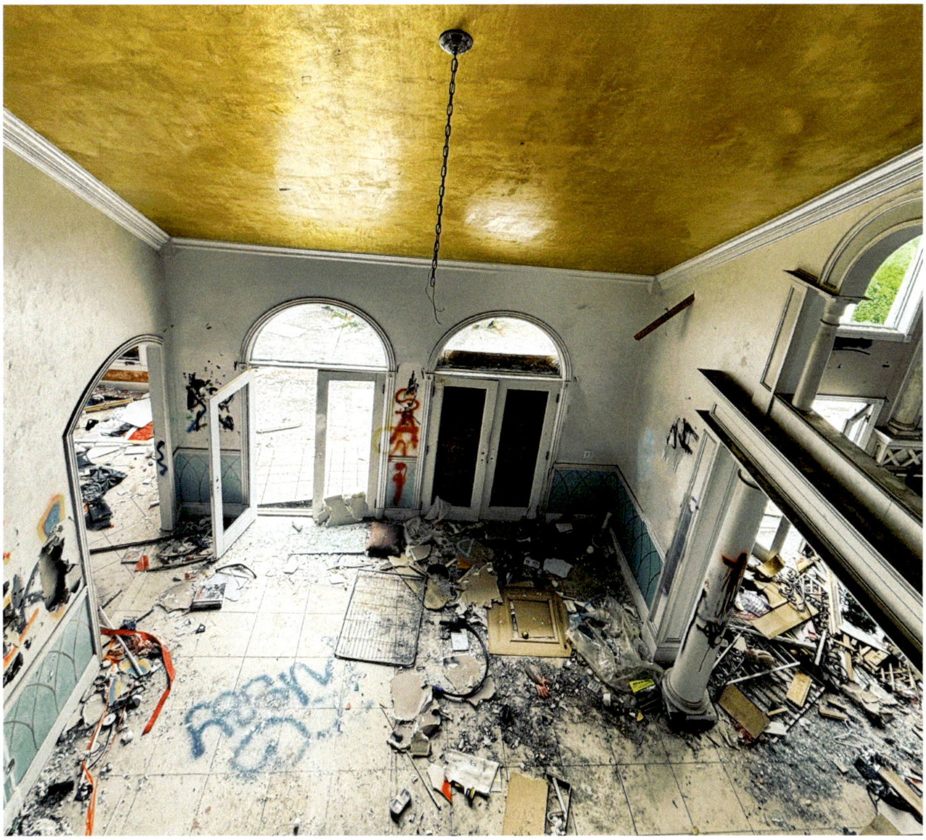

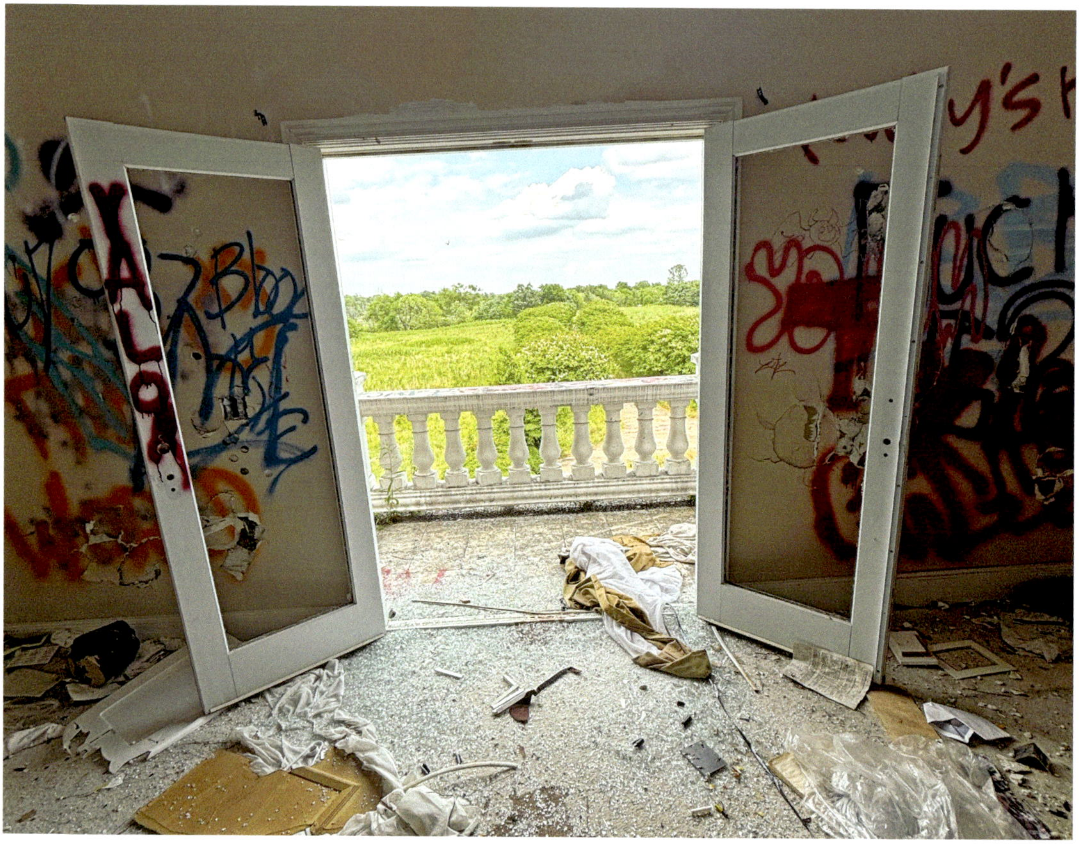

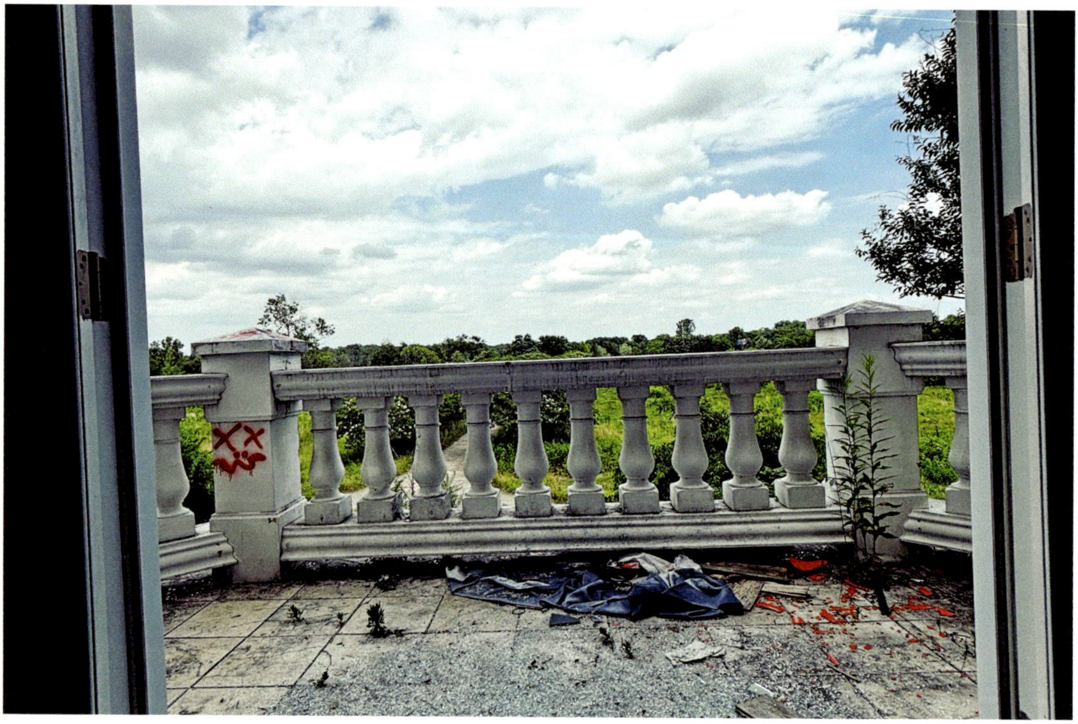

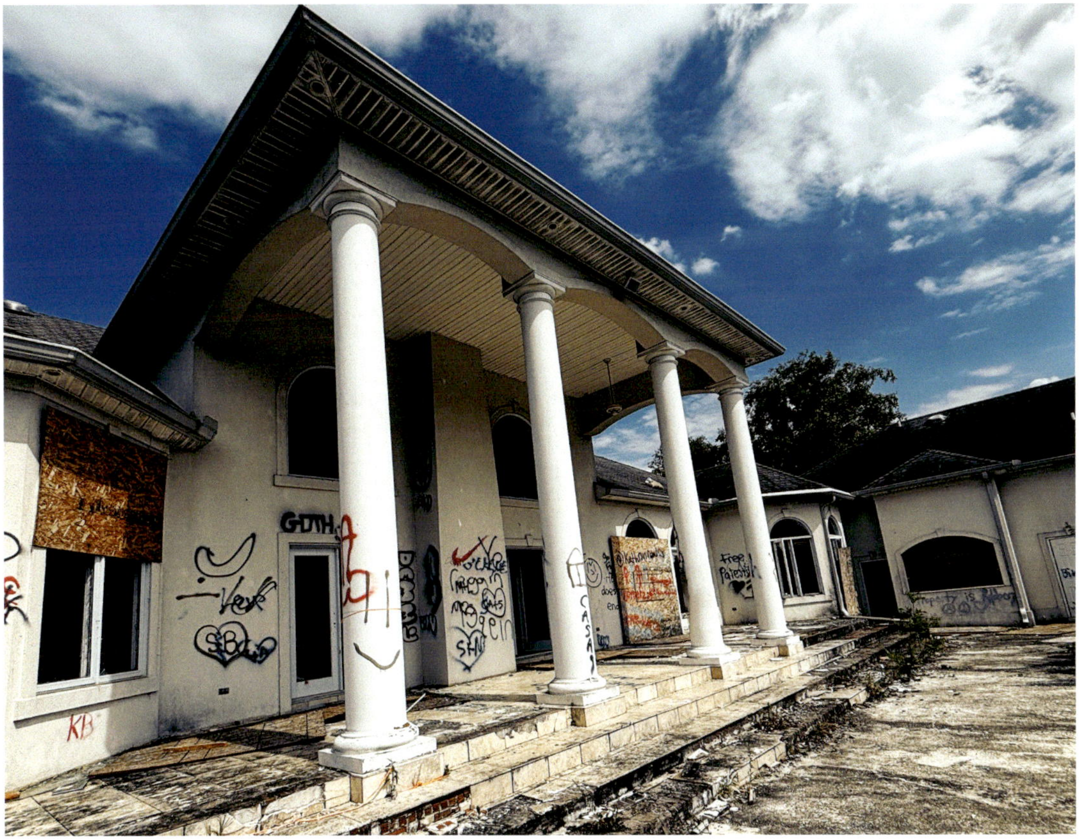

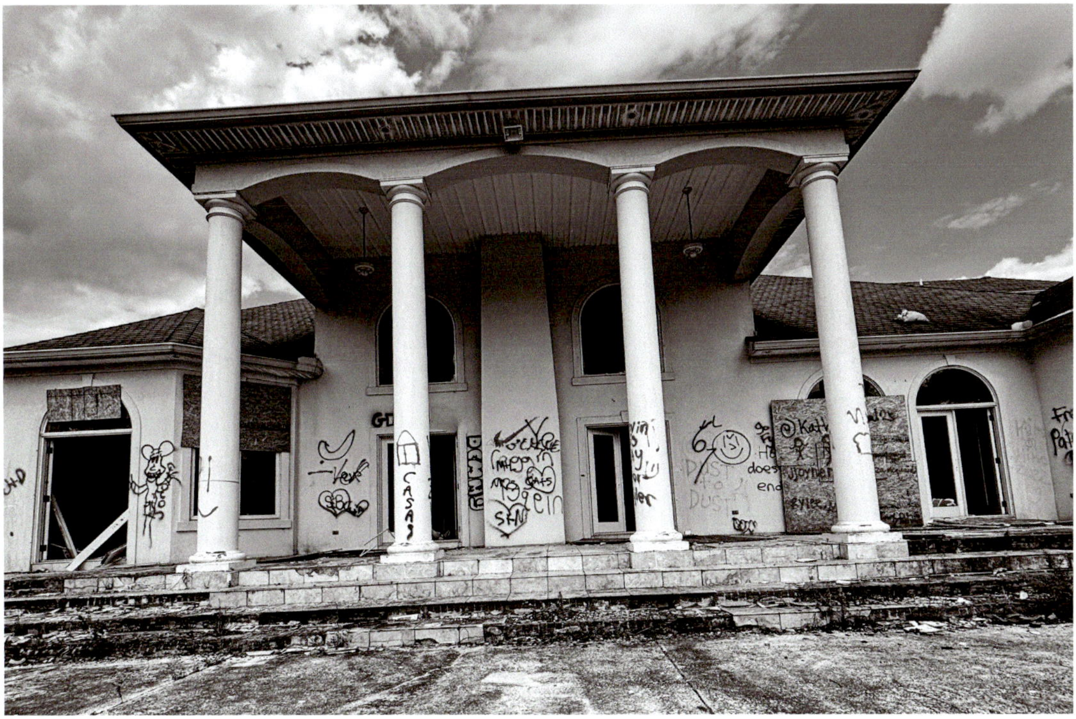

2

IT RAN WHEN I PARKED IT

I n forgotten barns, overgrown fields, and desolate scrapyards, a collection of antique and vintage vehicles quietly awaits rediscovery. Each photograph in this chapter tells the story of these abandoned relics, capturing their enduring beauty and melancholic charm.

These vehicles, remnants of bygone eras, stand as silent witnesses to the evolution of automotive design and craftsmanship. Their weathered surfaces, rusted bodies, and faded paintwork reveal the passage of time and the relentless embrace of nature reclaiming its territory. Yet, amid the decay, there is an undeniable allure—a nostalgic pull that invites us to reflect on the ingenuity and elegance of generations past.

From classic cars with their distinctive lines and chrome accents to rugged trucks and delicate carriages, each image unveils a piece of history frozen in neglect. Peering through shattered windows and rusted grilles, we glimpse interiors adorned with cracked leather seats, vintage gauges frozen in time, and steering wheels worn smooth by countless journeys.

These abandoned vehicles evoke a sense of mystery and intrigue, prompting us to ponder the stories they could tell. Who were their owners? What roads did they once travel, and what adventures did they witness? Each photograph invites us to explore not just the physical decay but also the enduring spirit of innovation and craftsmanship that defined an era.

As we embark on this visual journey through the world of abandoned antique and vintage vehicles, let us celebrate their enduring legacy and marvel at the beauty of these forgotten relics of the road.

Once a Bug, Now a Rug:
A Volkswagen Squareback's Farewell to the Open Road

I remember getting in my car at 5 a.m. to go find some hidden treasures to photo-graph during sunrise, and felt like it was my lucky day stumbling upon this driving down a state highway where I found the melancholic charm of a forgotten relic—a 1960s–1970s Volkswagen Squareback station wagon resting in the final throes of neglect. Its once-iconic cherry red paint now succumbs to rust, its chrome trim a faded memory. But a closer look reveals a more intriguing story. Body panels, like scattered puzzle pieces, lay sprawled inside, hinting at a valiant but ultimately unsuccessful attempt to save this once-loved companion. The weathered barn in the background offers a silent companion in decay, both remnants of a forgotten time where open roads and endless possibility beckoned.

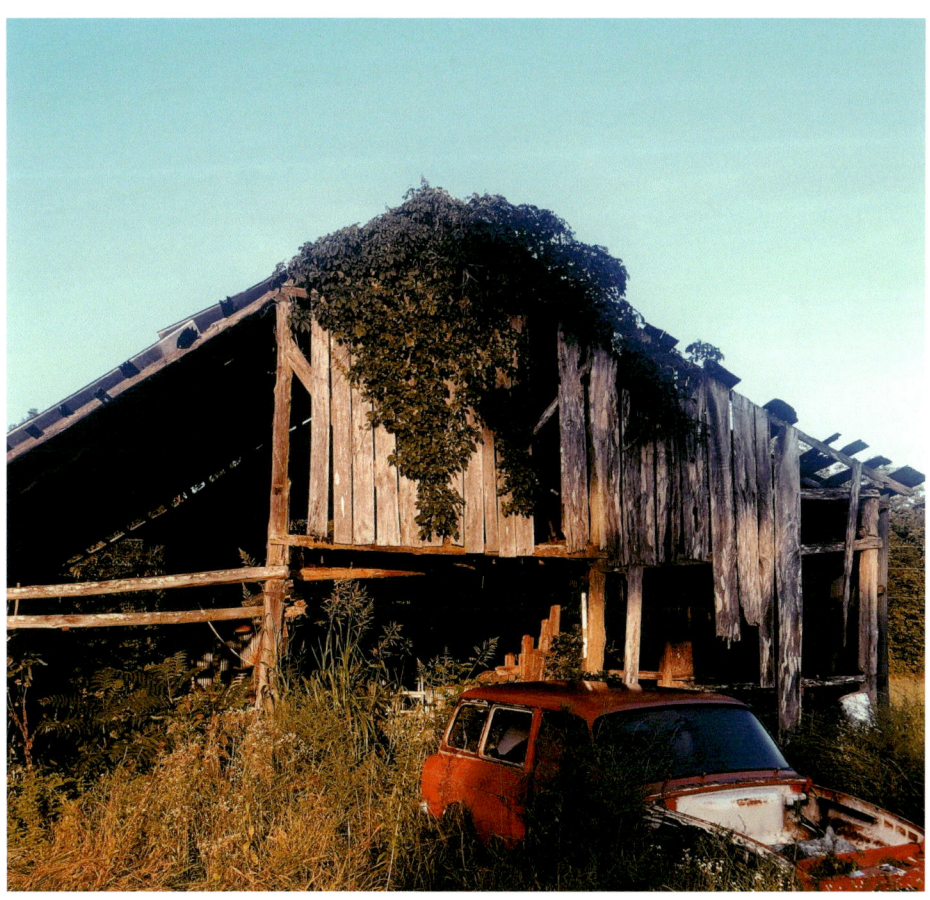

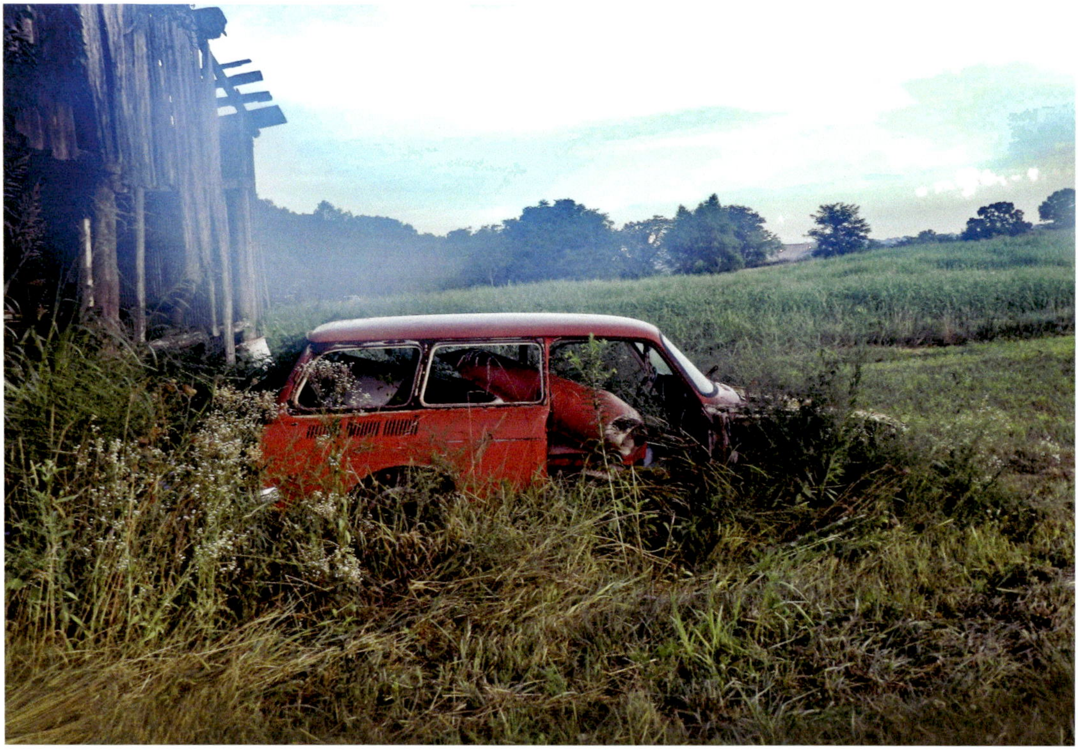

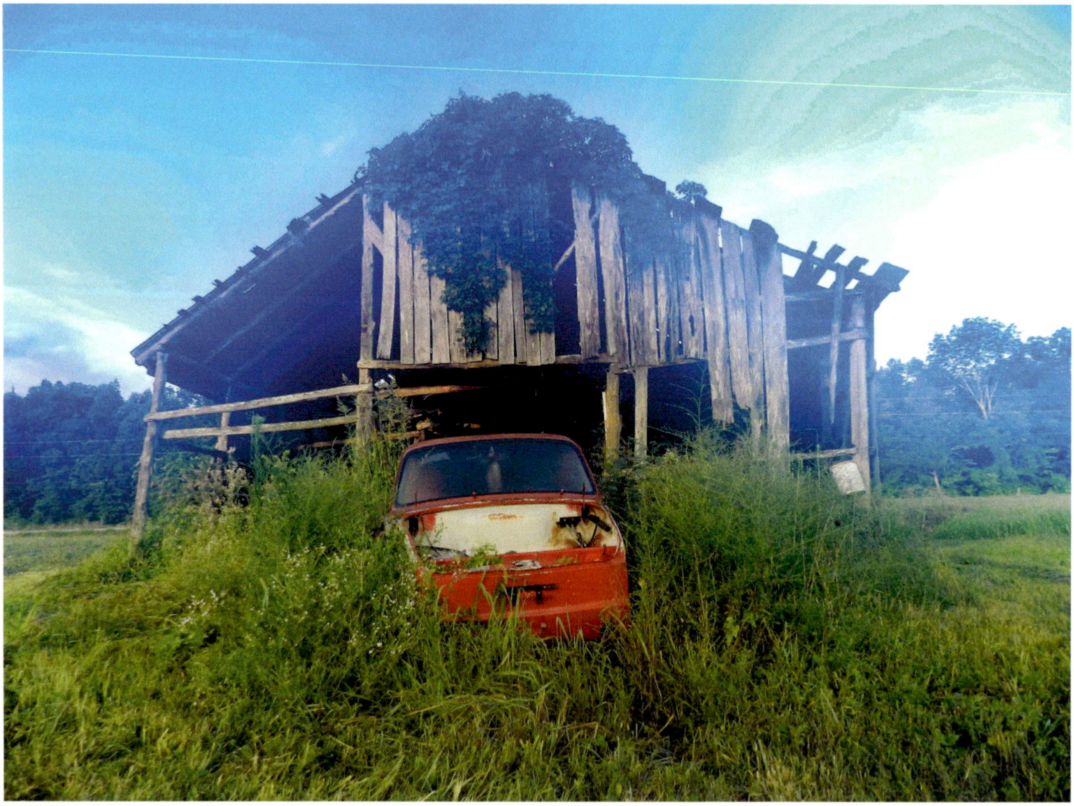

Chrome Dreams and Field of Green:
Where the '55 Dodge Went to Rest

Sun-baked and silent, a two-tone blue 1955 Dodge Custom Royal slumbers in a field of wildflowers, its final resting place. A rusted license plate, a whisper of a forgotten identity, peeks through the weeds. This beauty, once a head-turner with gleaming chrome and aerodynamic curves slicing through the air, begs the question: who was the driver, the one who commanded attention on the open road? Back then, cars were built to last, not just aesthetically—their solid build and mechanical ingenuity promised adventures untold. Now, nature reclaims its canvas, transforming this steel stallion into a time capsule. These photos capture the fading glory of a bygone era, a testament to a time when machines were crafted with both brawn and beauty.

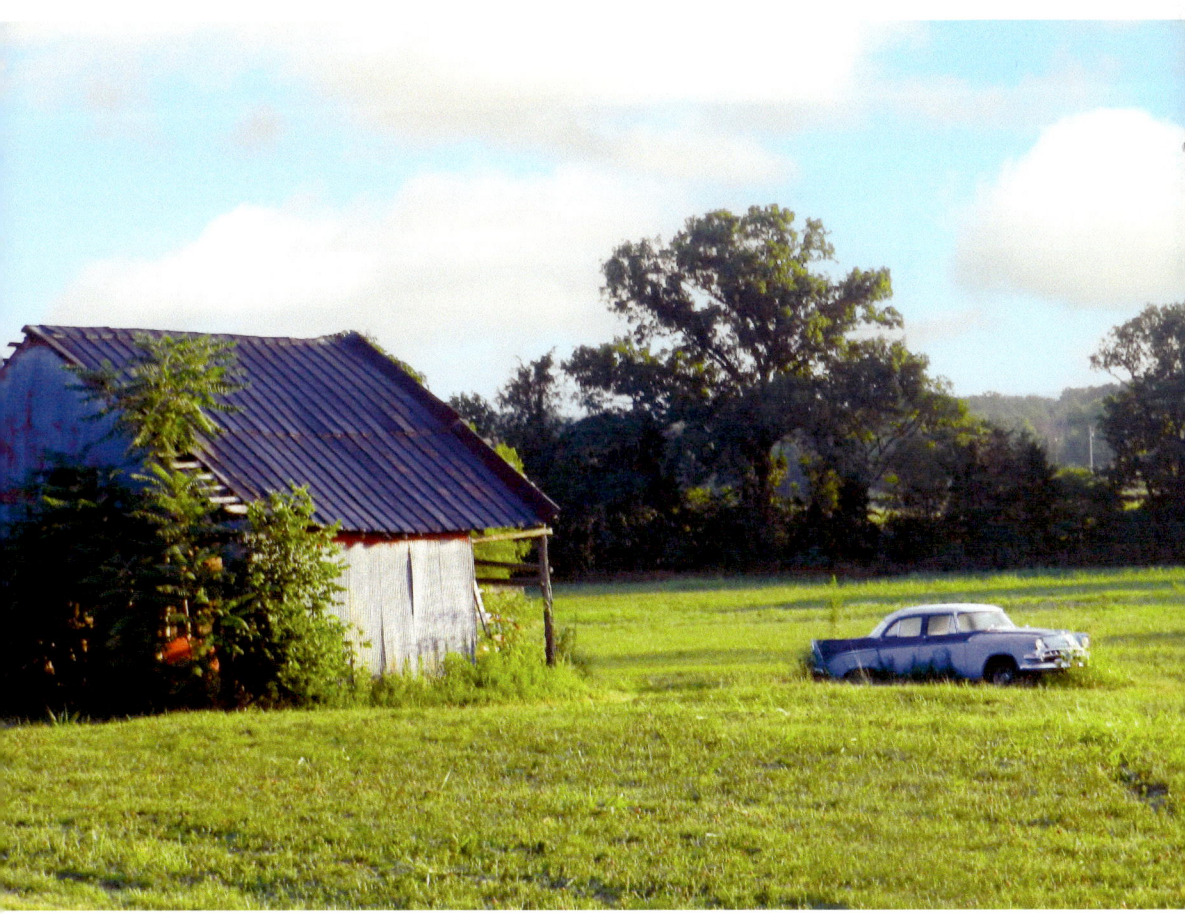

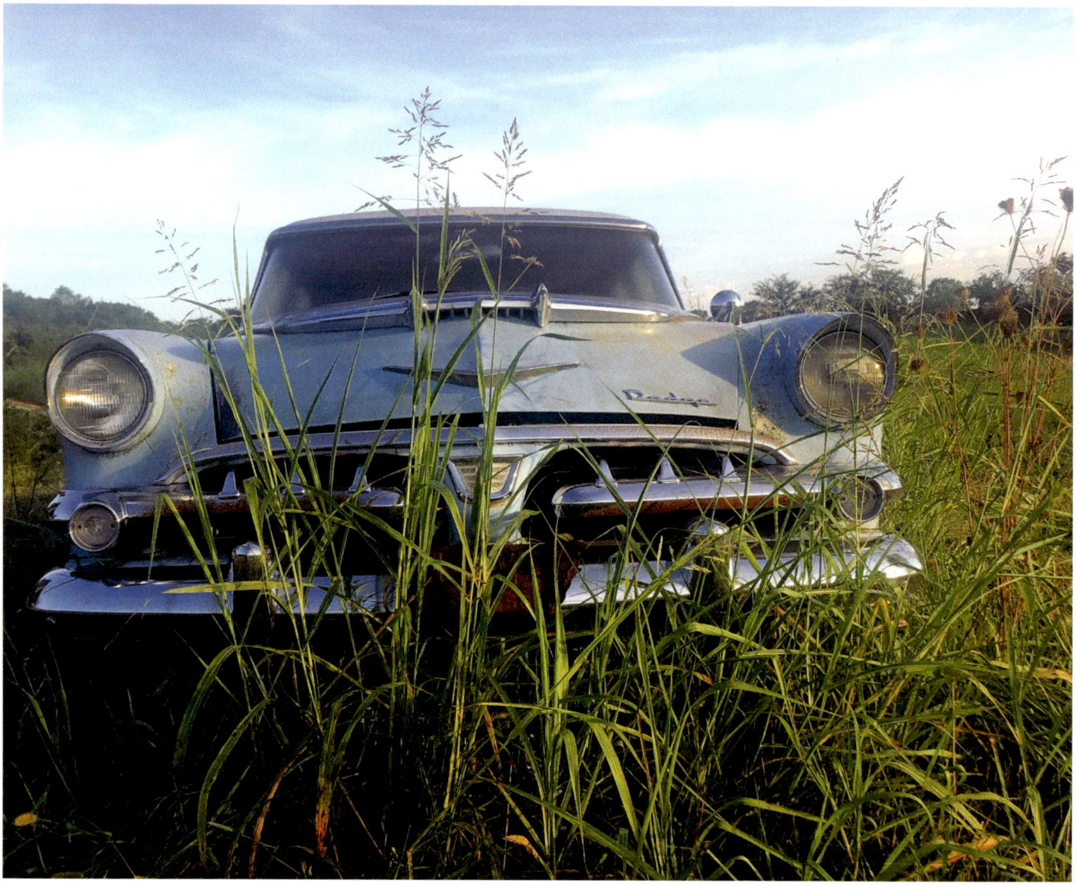

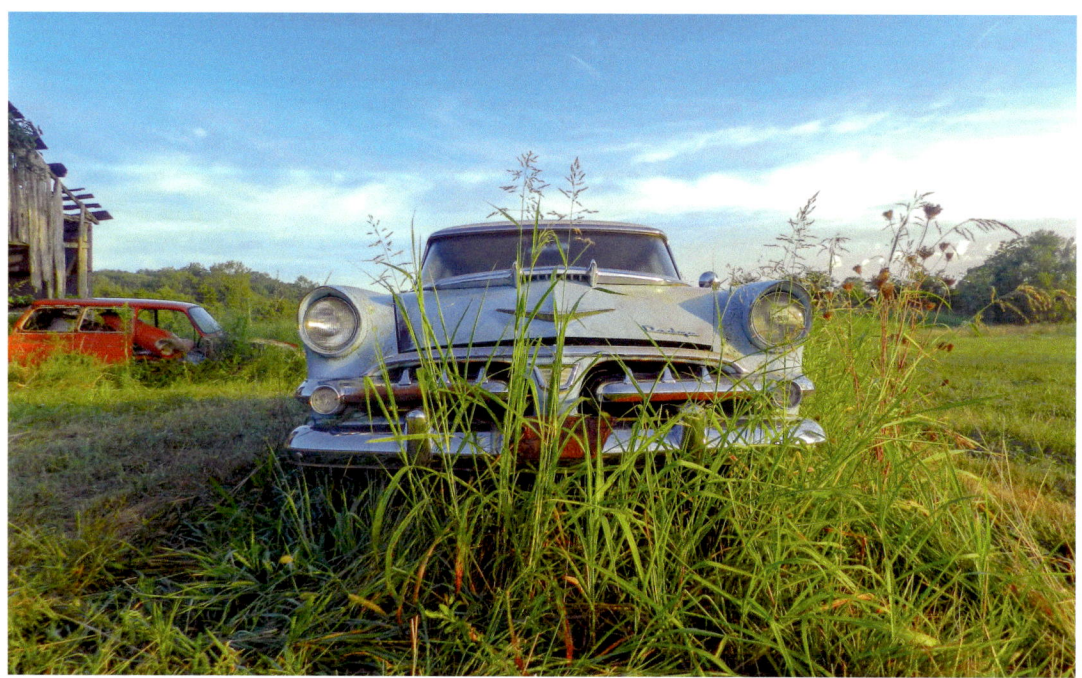

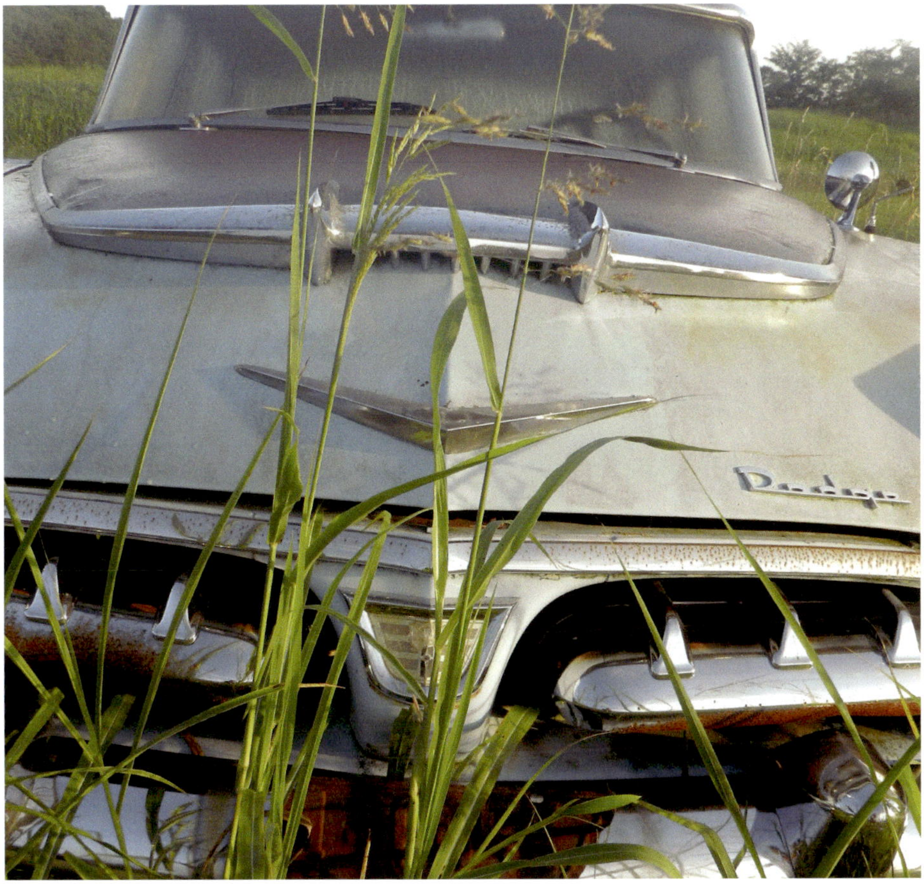

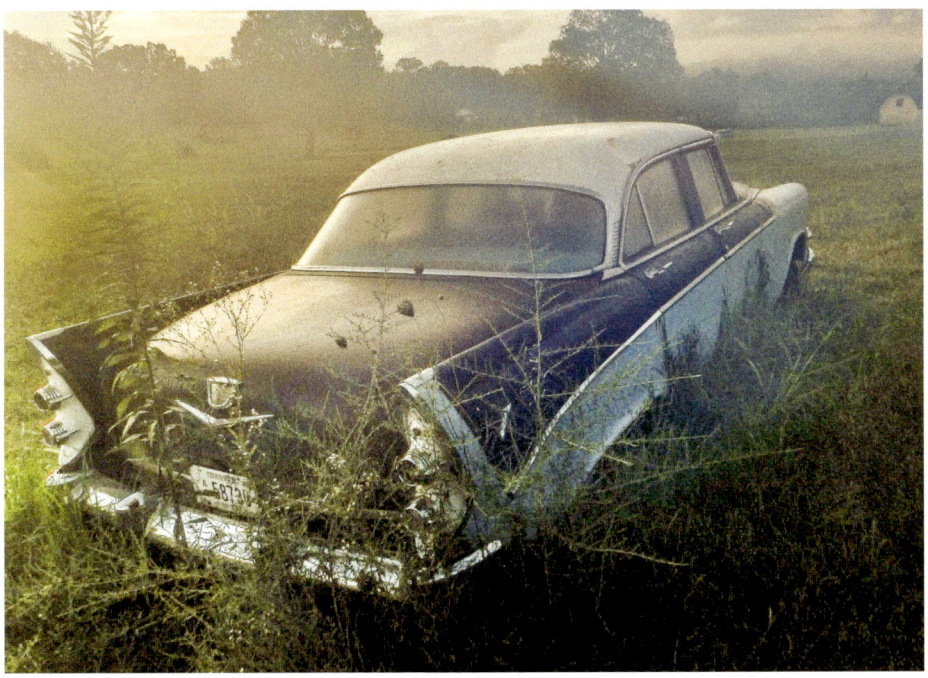

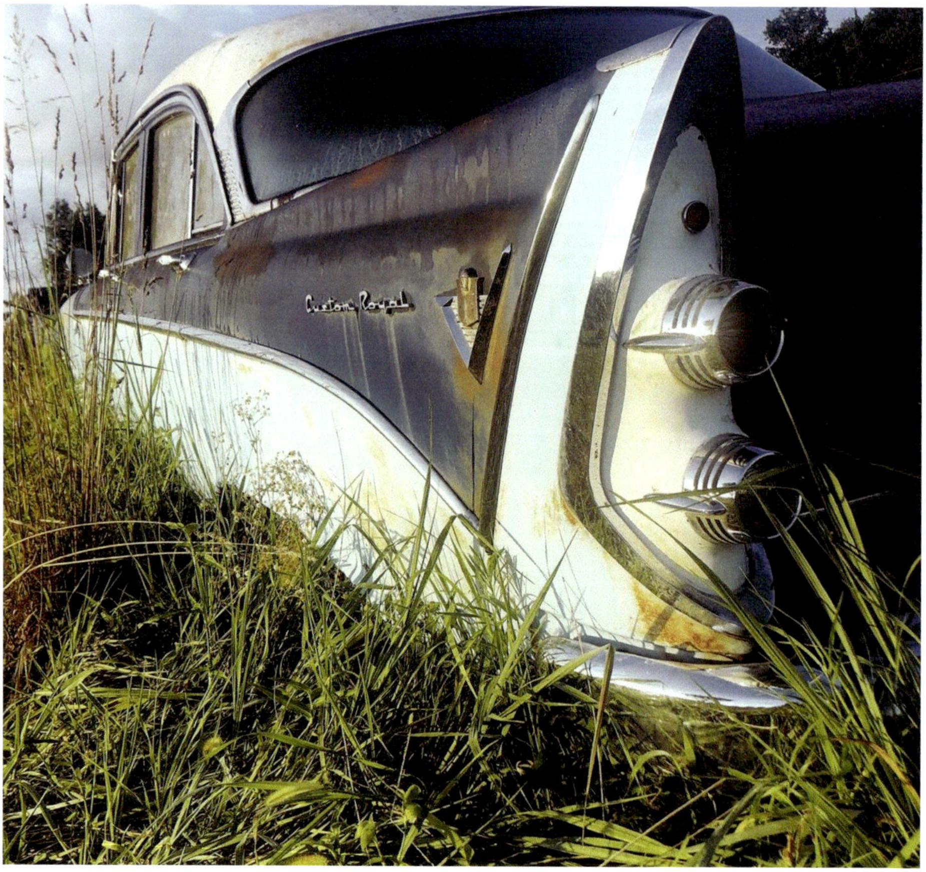

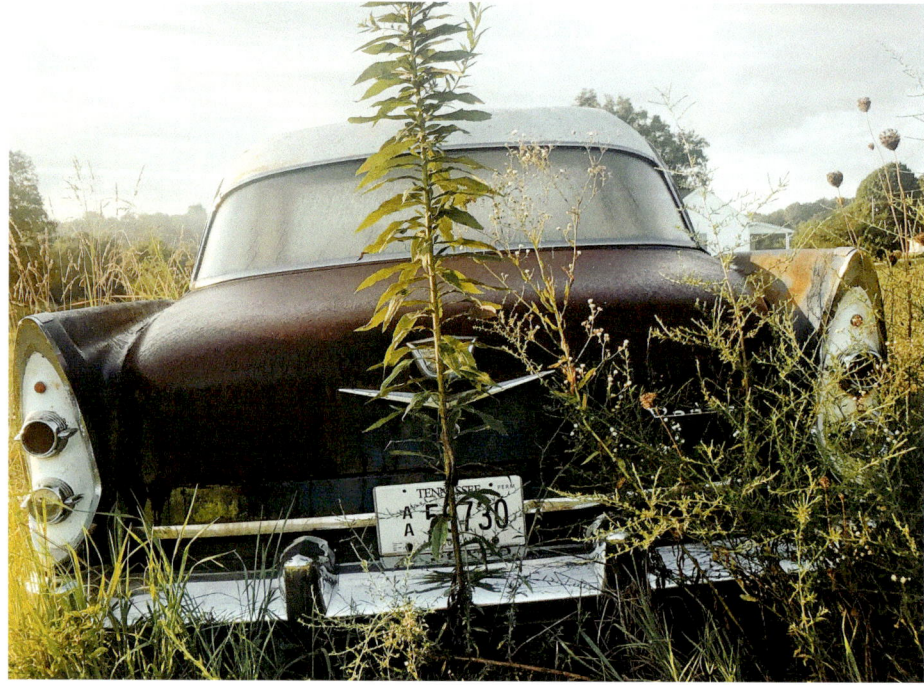

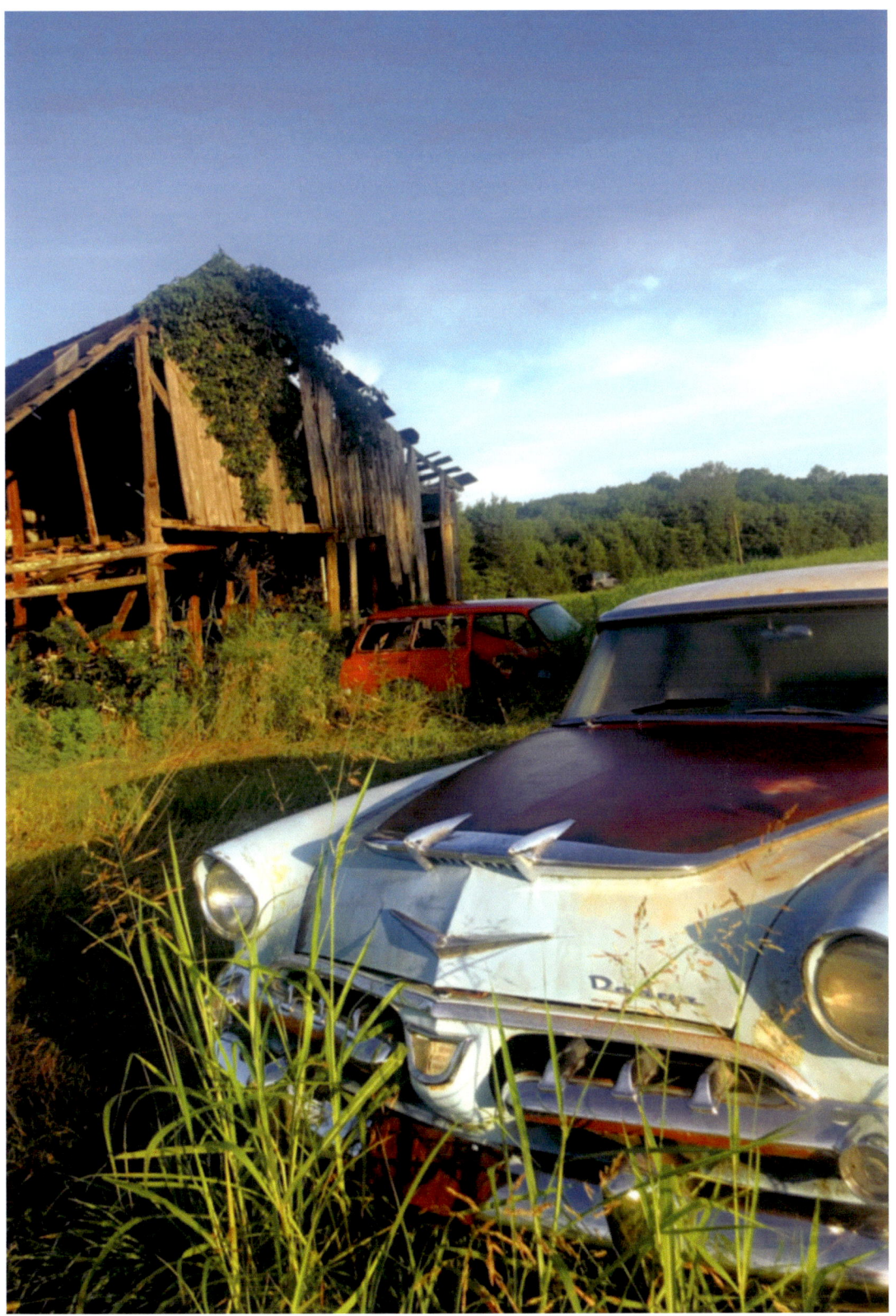

The Forgotten Freight: A Boxcars Unsent Journey

Among the country roads, near a wooded area, is a forgotten sentinel—a weathered boxcar slumbering in a field of tall grass. Rust has devoured its once-vibrant paint, leaving behind a cryptic message—faded words hinting at a journey long since forgotten. Its missing doors spill secrets into the sunlight—a jumble of miscellaneous items, remnants of a life lived on the move. Was it a home, a temporary haven, or simply a vessel for forgotten cargo? The mystery lingers, just like the boxcar itself, a silent monument to journeys past.

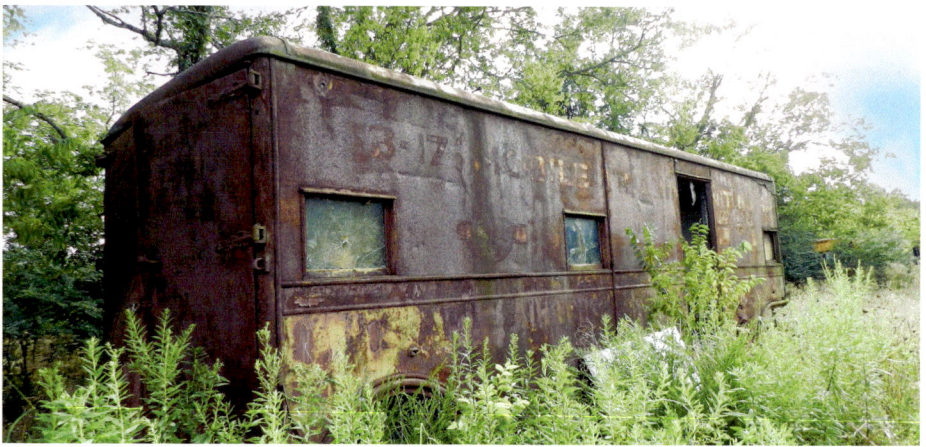

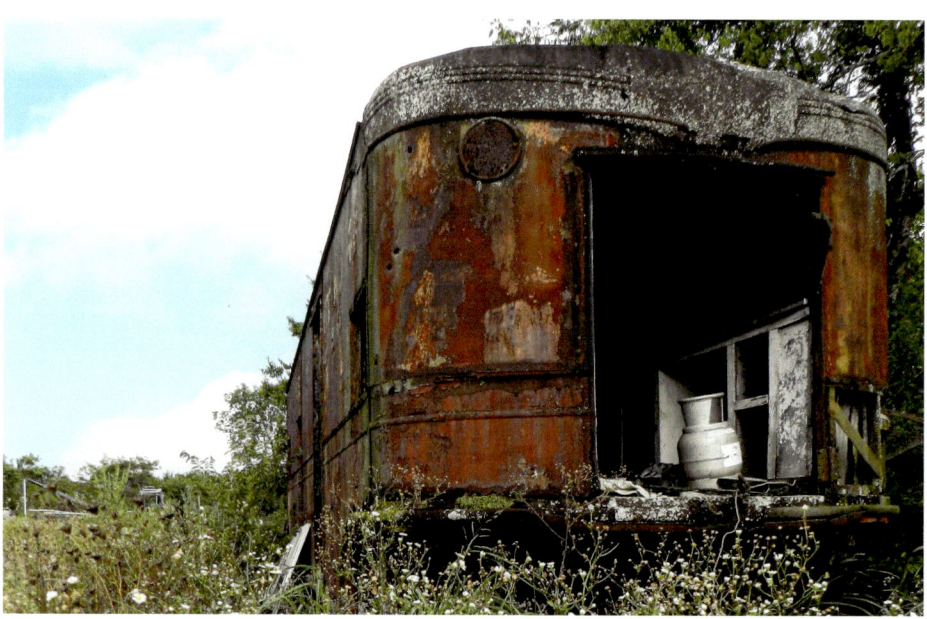

The Last Stop: A School Bus's Forgotten Journey

Resting in an overgrown field along a backroad, an abandoned school bus hid behind a fallen brick structure and from the roadway, I saw the melancholic charm of a forgotten field trip—a once-vibrant school bus now a lonely sentinel in a sea of tall grass. Its once-bright paint, at one time a beacon for eager students, now fades under the relentless sun. Rows of empty seats whisper of forgotten laughter and childhood dreams. Was it a mechanical failure, a budget cut, or a school that closed that left this bus stranded in this field? Each photo offers a glimpse into this steel time capsule, a silent testament to the passage of time and the enduring power of childhood memories.

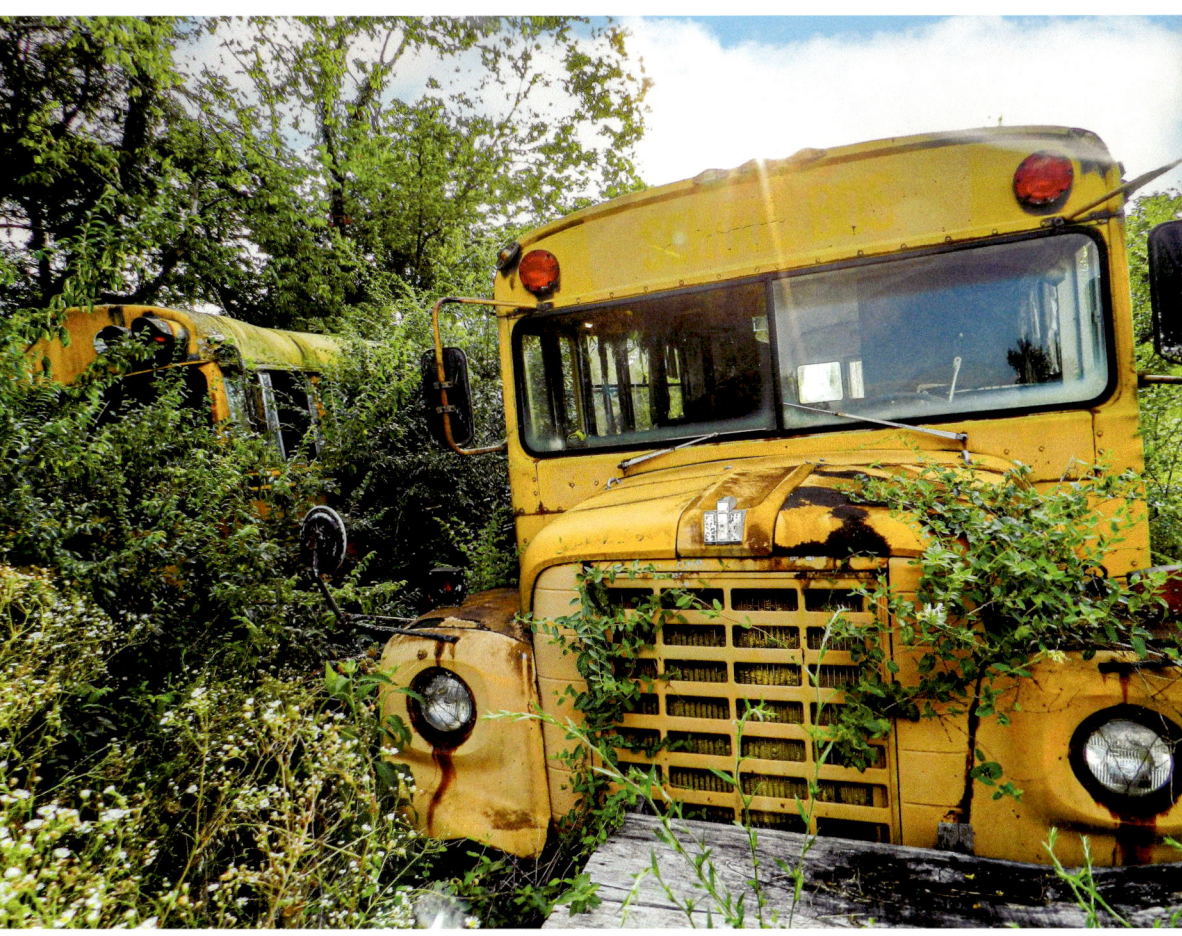

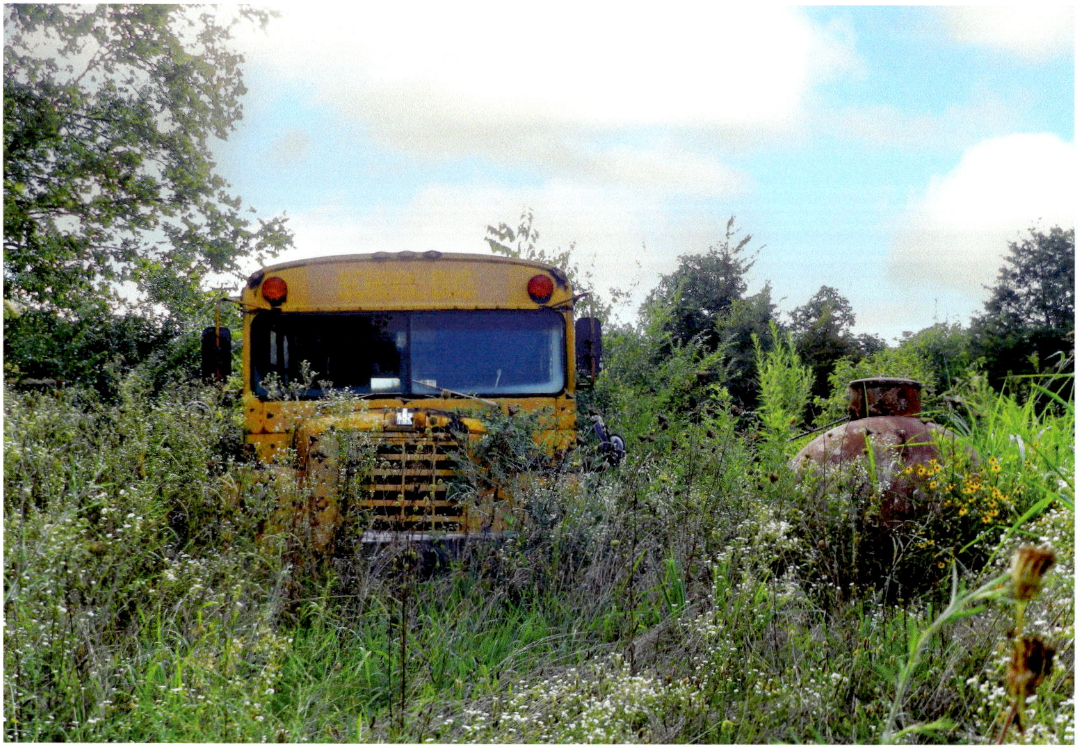

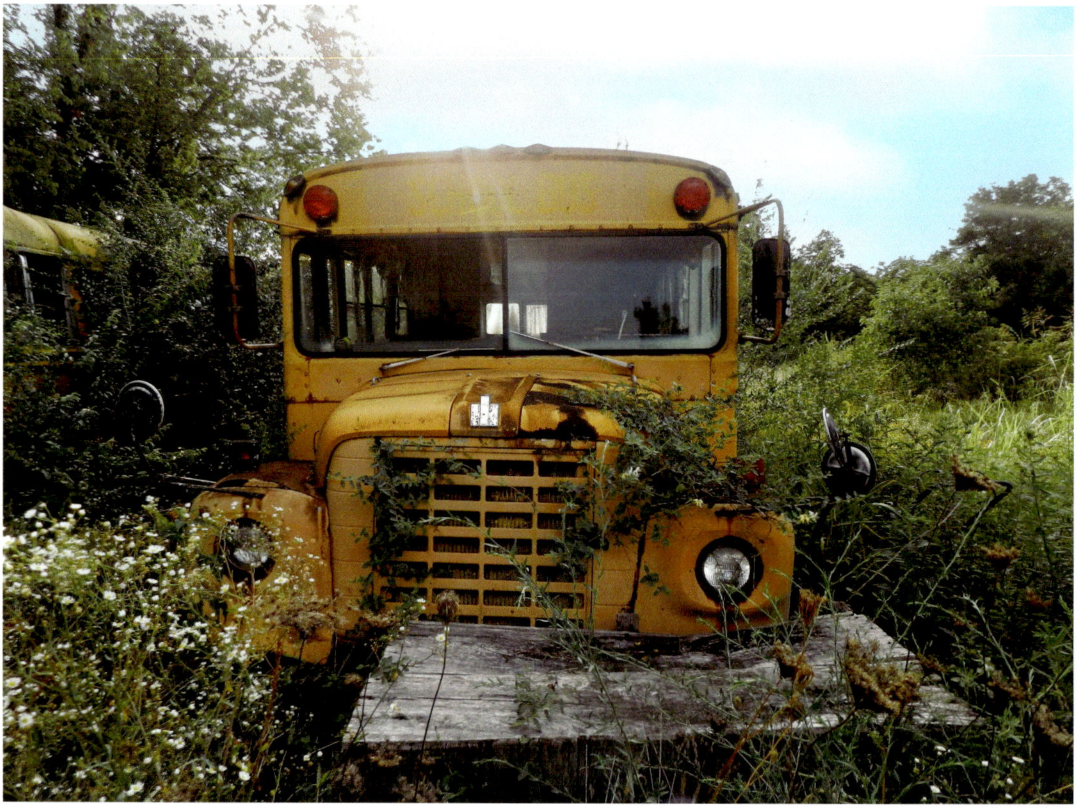

Sun-Bleached Smile: A Chevy Grille's Faded Greeting

Sun-bleached Chevy grille, a relic in a wildflower field. Faded chrome whispers of a forgotten journey. What classic car did it adorn? The silence holds only echoes of an era's open road.

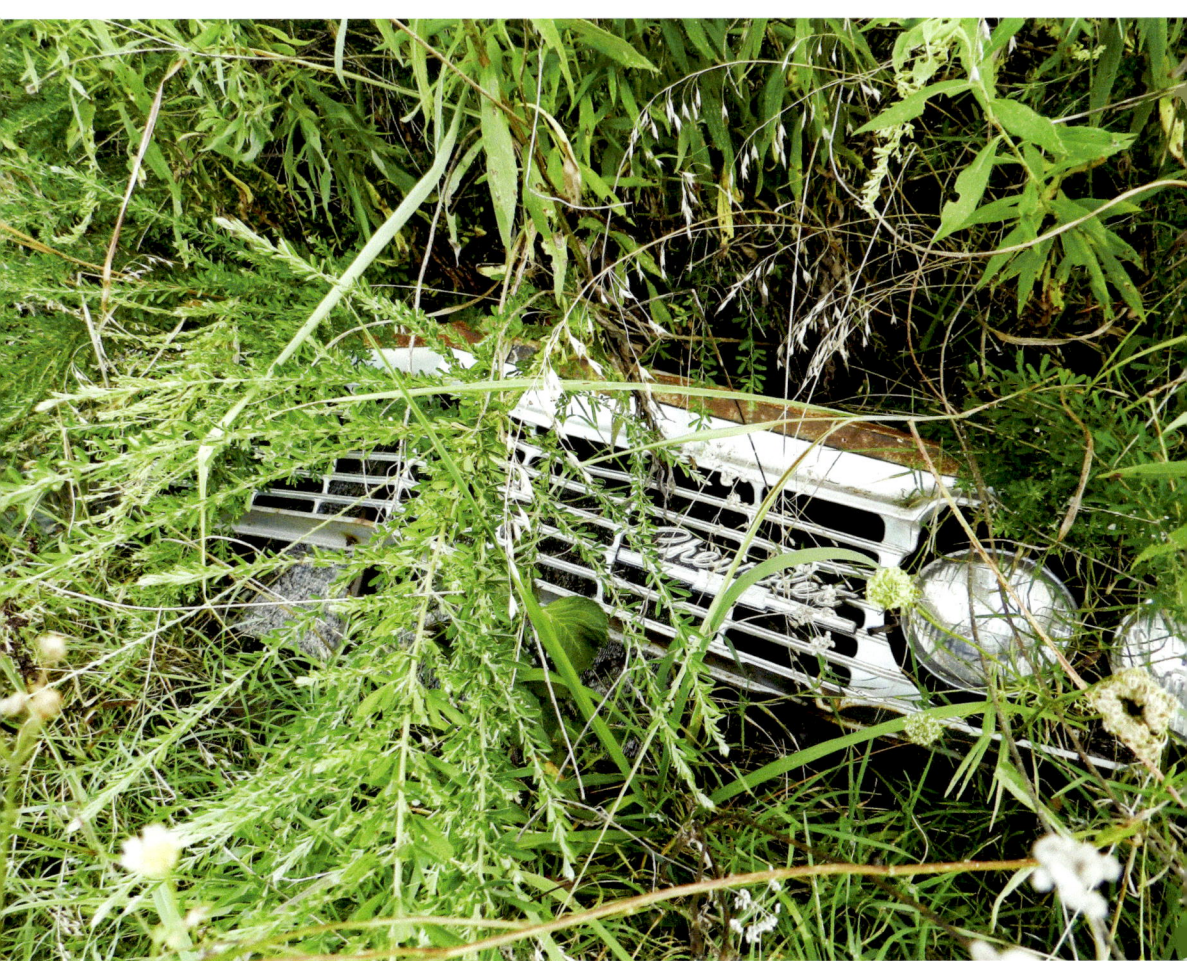

Abandoned Workhorse: A Ford F-Series Forgotten Task

Driving through Readyville early one morning, I spotted a lone warrior—a 1976 Ford F-Series Truck, now a solitary sentinel in a field swallowed by nature. The faded blue paint, once gleaming with pride, hints at countless adventures. Was it a farmer's faithful companion, hauling countless harvests? Or perhaps a hunter's trusted steed, conquering dirt roads in pursuit of the perfect trophy? We imagine the owner, a grin stretching across their face, keys clutched in hand, the day they first drove this beauty off the lot. Now, the silence is deafening. Vines snake through the open windows, and the once-sturdy bed is a canvas for wildflowers. The rust whispers tales of forgotten journeys, leaving us to wonder what brought this loyal workhorse to its final resting place. Was it a heart-wrenching breakdown, a well-deserved retirement, or something more? The mystery lingers, only adding to the melancholic charm of this forgotten Ford, a testament to a time when machines were built to last and adventures knew no bounds.

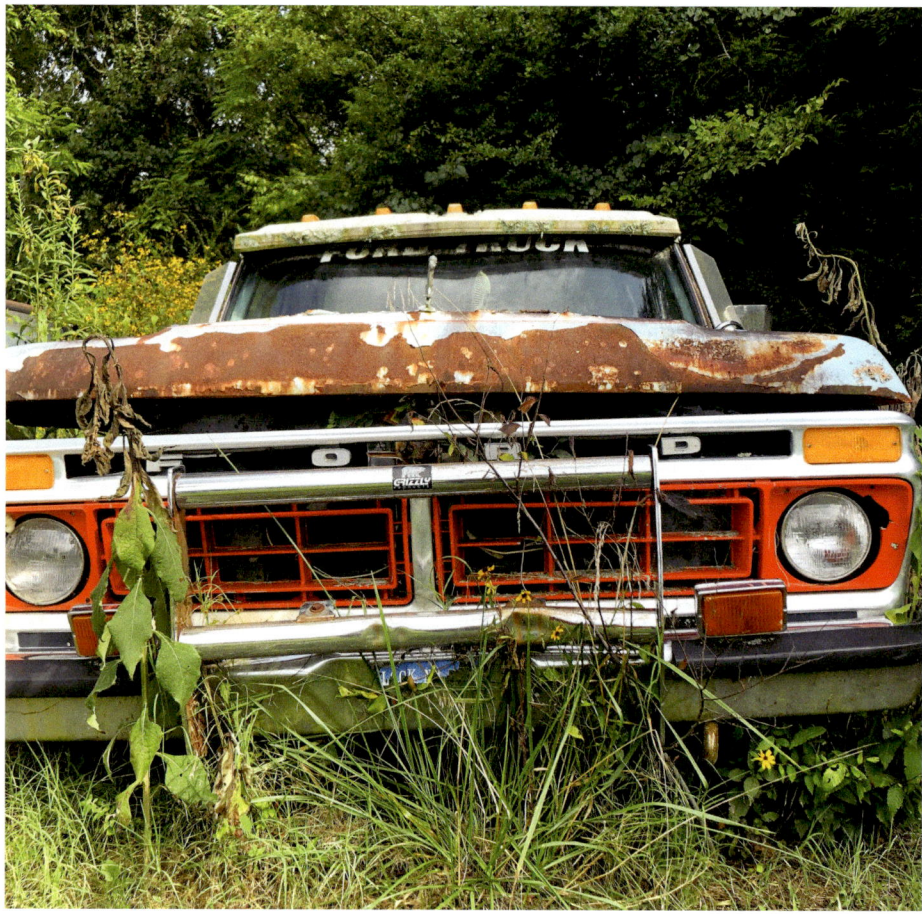

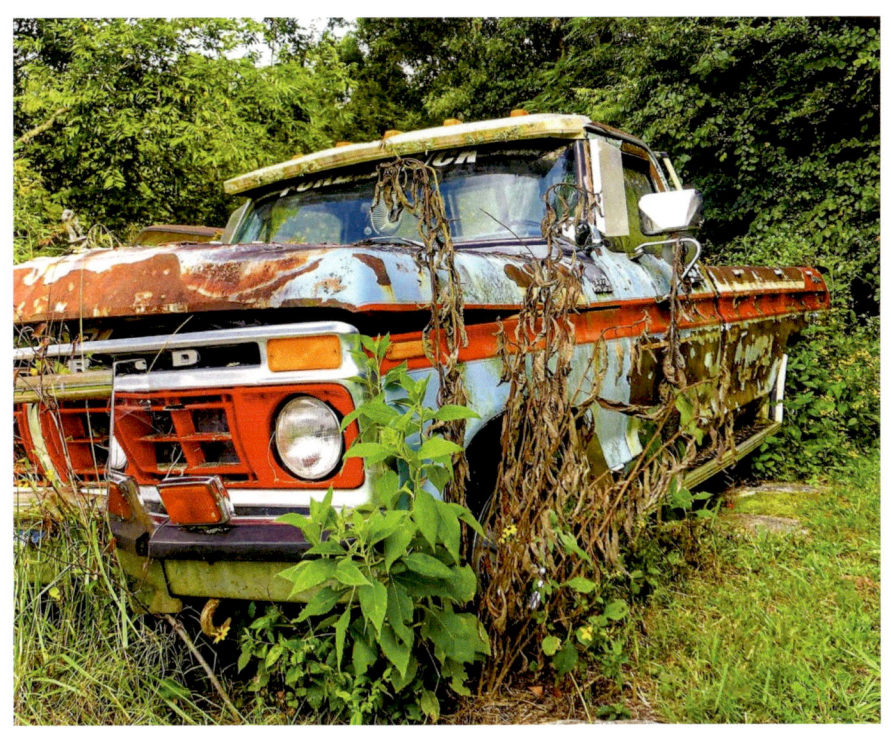

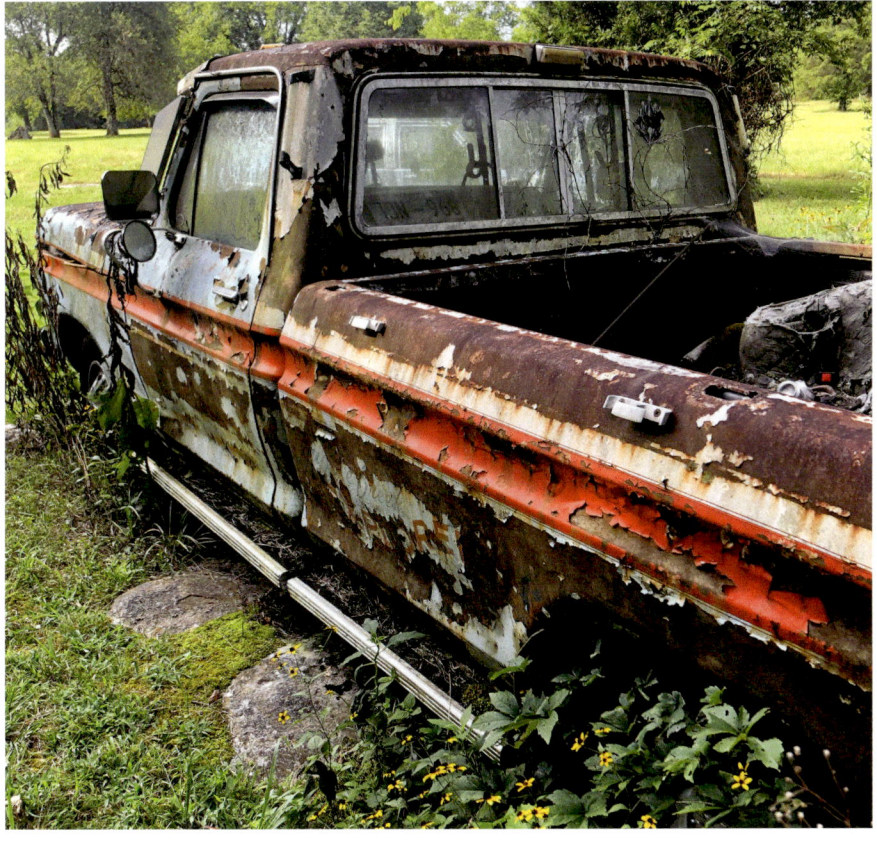

From Worksite to Wildflowers: A Chevy C/K's Second Bloom

Along the backroads of Readyville early one morning, the sun was rising, and I spotted a 1970s–1980s Chevrolet C/K truck, slumbering in a field reclaimed by nature. The once-gleaming chrome is now a dull echo of its former glory. We imagine the sunbaked days spent hauling, the mud-caked nights pushing through backroads. Was this a farmer's confidant, its bed overflowing with golden harvests? Or perhaps a builder's partner, its sturdy frame bearing the weight of dreams taking shape? The tall grass shrouds its final chapter in mystery. Vines grasp at the rusting frame, a silent testament to the relentless passage of time. This Chevy C/K, a monument to an era of American grit, stands as a reminder that even the sturdiest machines eventually succumb to nature's embrace.

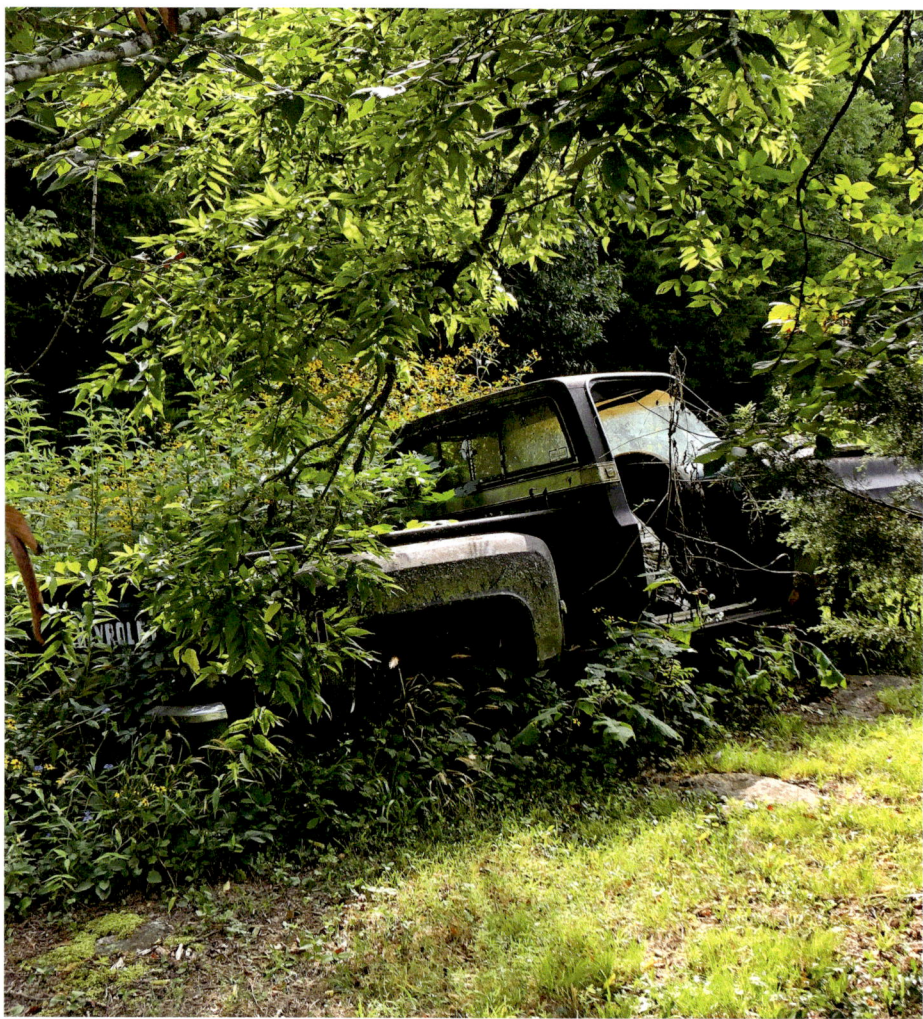

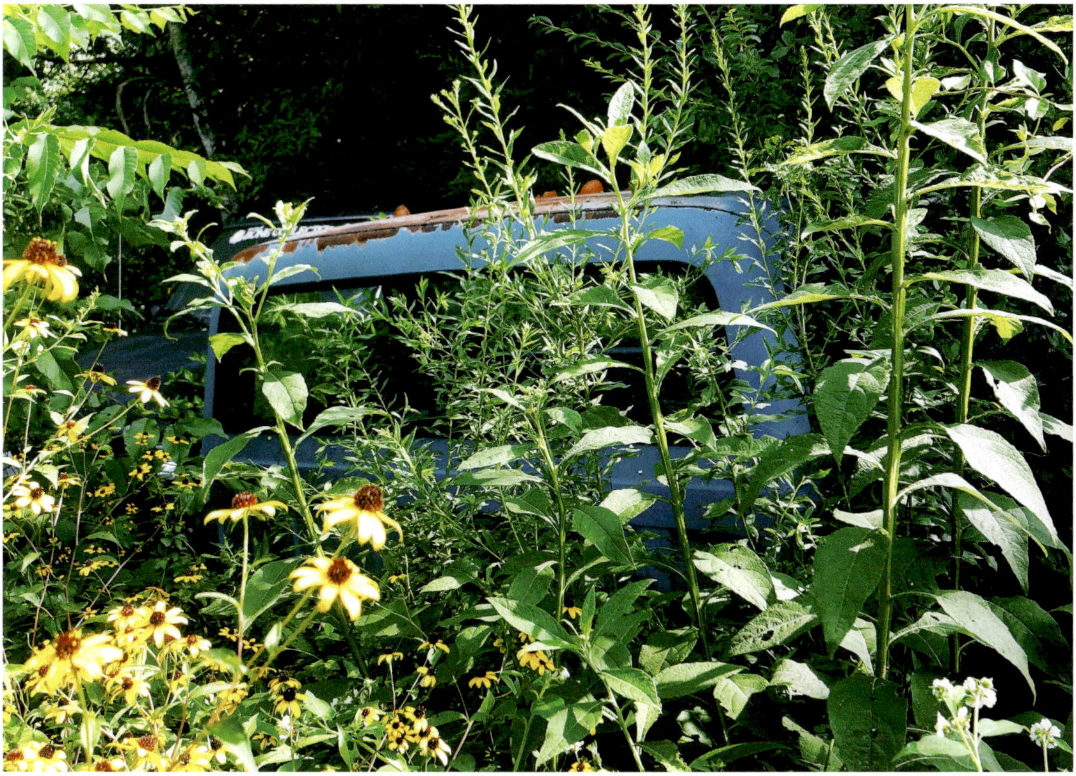

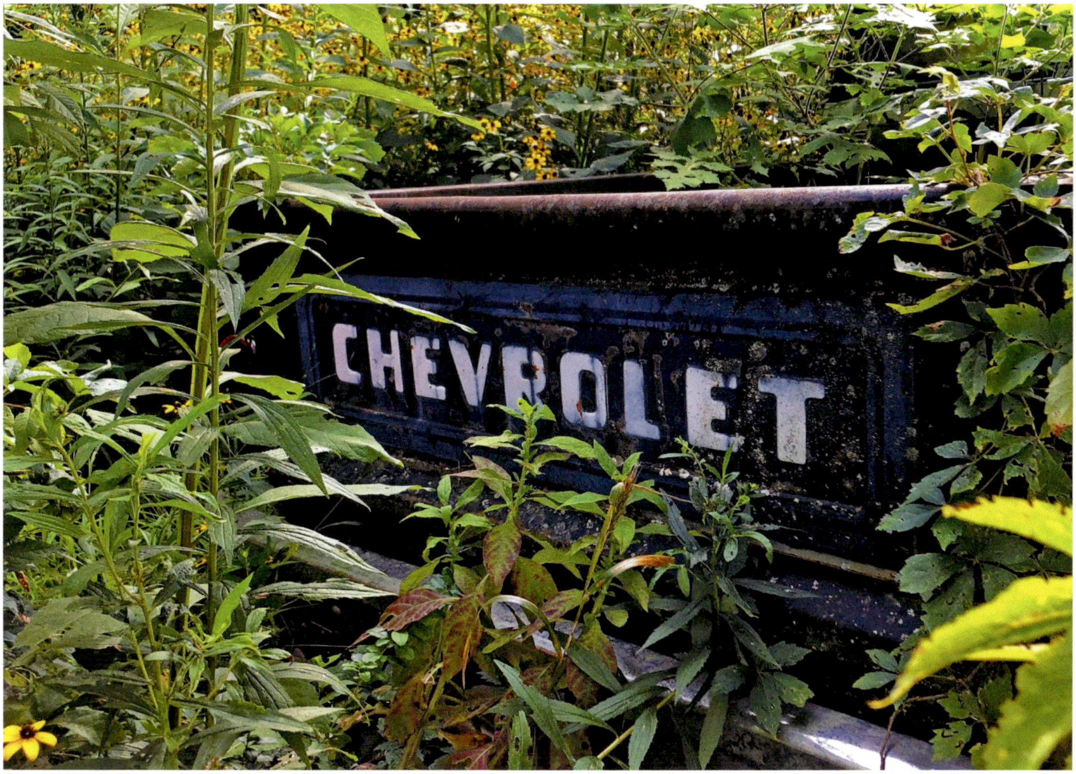
CHEVROLET

From Mud Bog to Meadows: A Jeep CJ's Final Rest

Sunbaked Jeep with the key still in the ignition, a forgotten adventure. Vines climb mud-caked fenders, whispers of past trails. Did a driver flee, or did fate intervene? The Jeep slumbers, a monument to exploration, its spirit urging us to seek our own off-road journeys.

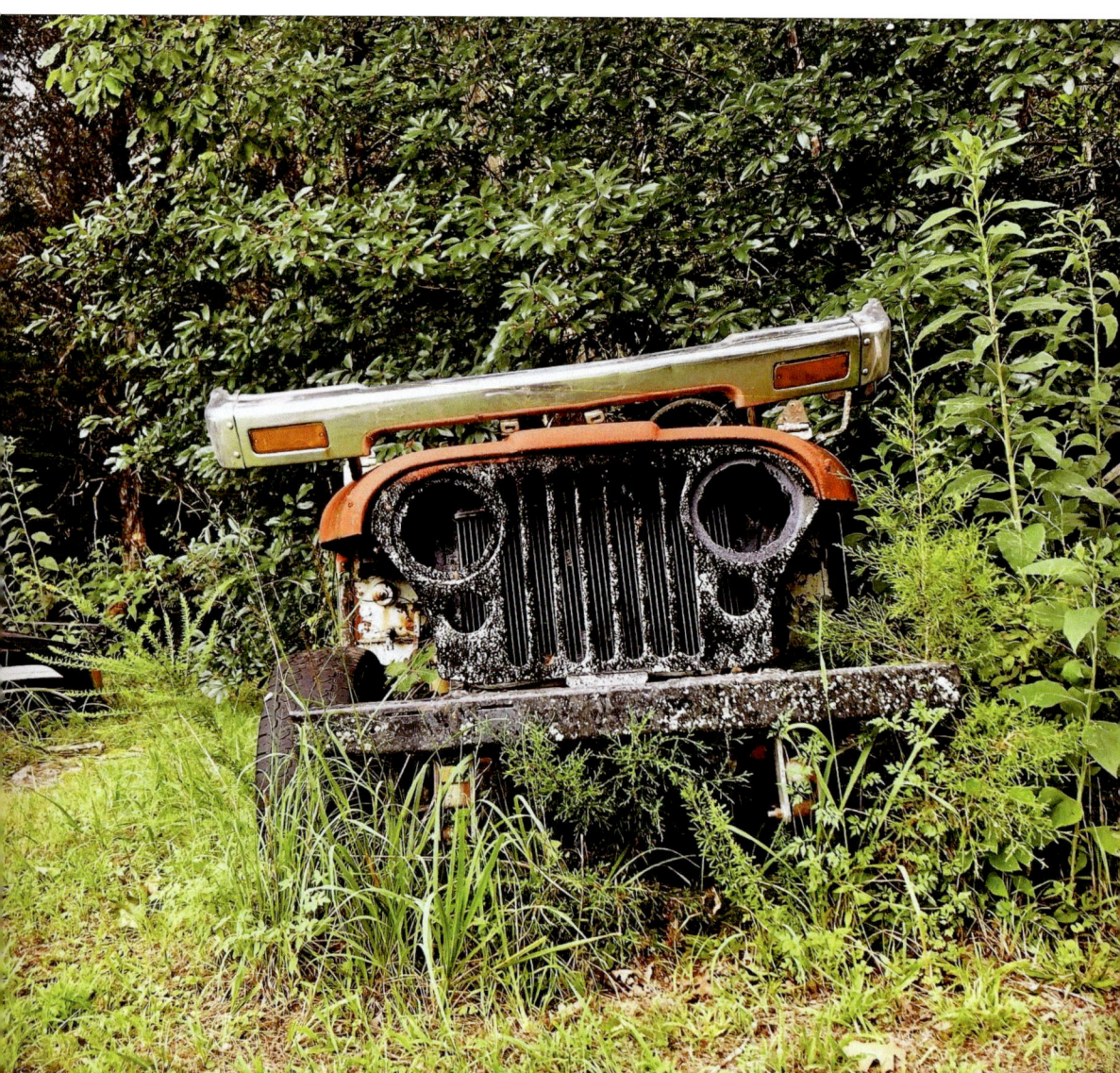

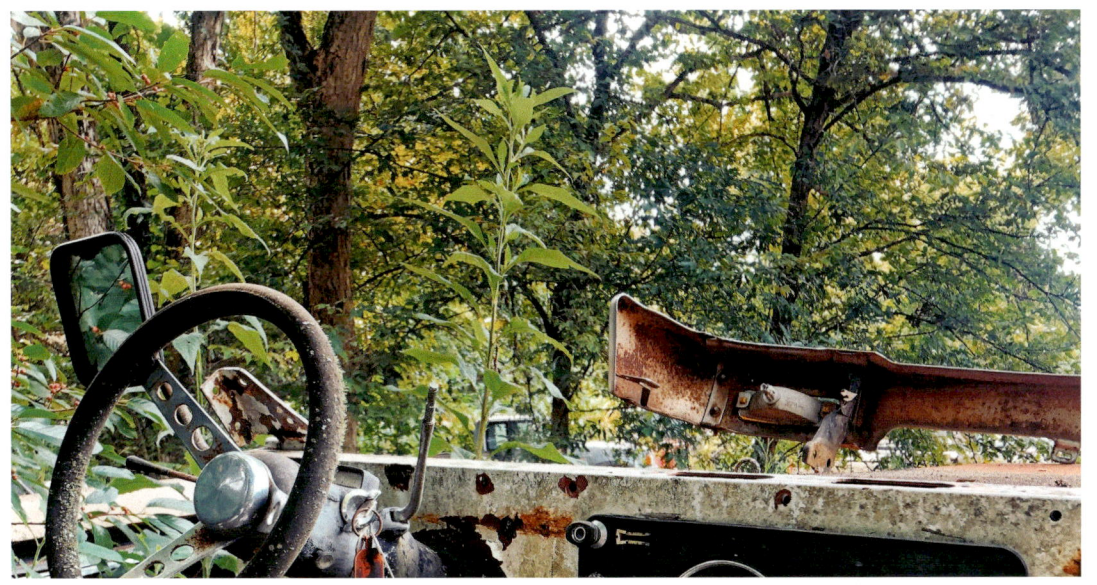

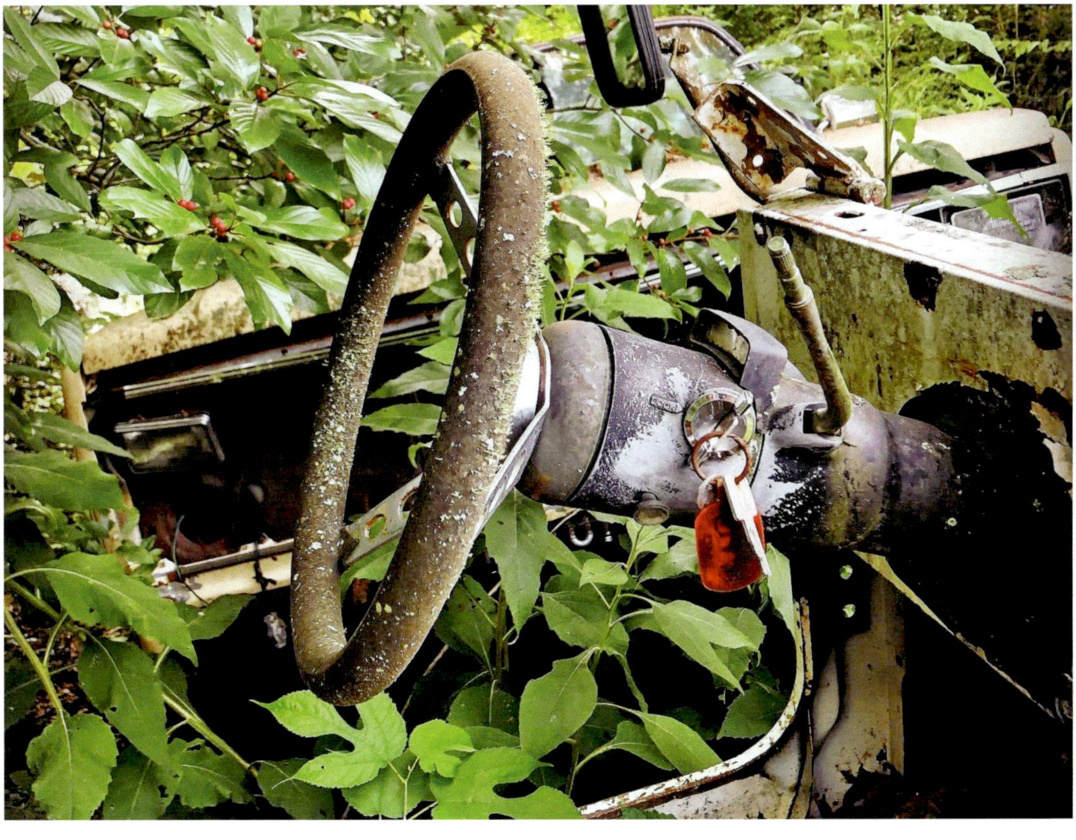

3

RUSTED RELICS

I n the communities that surround Middle Tennessee, history lingers like a humid
breeze, forgotten artifacts of daily life lie scattered across the landscape, hidden
in plain sight. This chapter takes you on a visual journey through the hauntingly
beautiful remnants of abandoned gas stations, grocery stores, and farmland. These
are not just places, but echoes of lives lived, now immortalized through the art of
urban exploration photography.

Amid the rusting skeletons of gas pumps, you can almost hear the murmur of
travelers who once paused here, their lives intersecting for a brief moment before
moving on. Dust-covered aisles in long-deserted grocery stores evoke a time when
these shelves brimmed with goods, serving the local community whose vibrancy
has since faded into silence. And on sprawling farmlands, weather-beaten barns
and obsolete machinery stand as stoic witnesses to the relentless march of time,
each telling its own silent tale of labor, hope, and abandonment.

This chapter is a tribute to these forgotten corners of the South, where decay
intertwines with nostalgia, creating a beauty that speaks to the impermanence of
human endeavor. The photographs within this chapter invite you to step into these
neglected spaces, to feel the weight of history, and to imagine the lives that once
animated these now-still environments.

Through the lens of urbex photography, we delve into the soul of these deserted
places, capturing not just the physical decay but the essence of what remains.
Prepare to be moved by the stark yet captivating images of the South's rusted relics,
where each frame is a window into a past that is both distant and achingly familiar.

From Convenience to Cravings: A Corner Store's Delicious Rebirth

This collection of photos taken in Rockvale charts the delightful transformation of a small-town landmark. Once a beloved corner store, its shelves stood empty, echoing with memories of community gatherings and forgotten errands. But thanks to the vision of new owners, this space has been reborn as a thriving country cafe. Step through the doors, and subtle nods to the store's past peek through. Each photo captures the essence of this metamorphosis, a testament to both progress and preservation. The aroma of freshly brewed sweet tea and country cooking mingles with the whispers of history, creating a warm and inviting atmosphere. This cafe is not just about delicious food; it is a celebration of the community's spirit, ensuring the corner store's legacy lives on, one satisfied customer at a time.

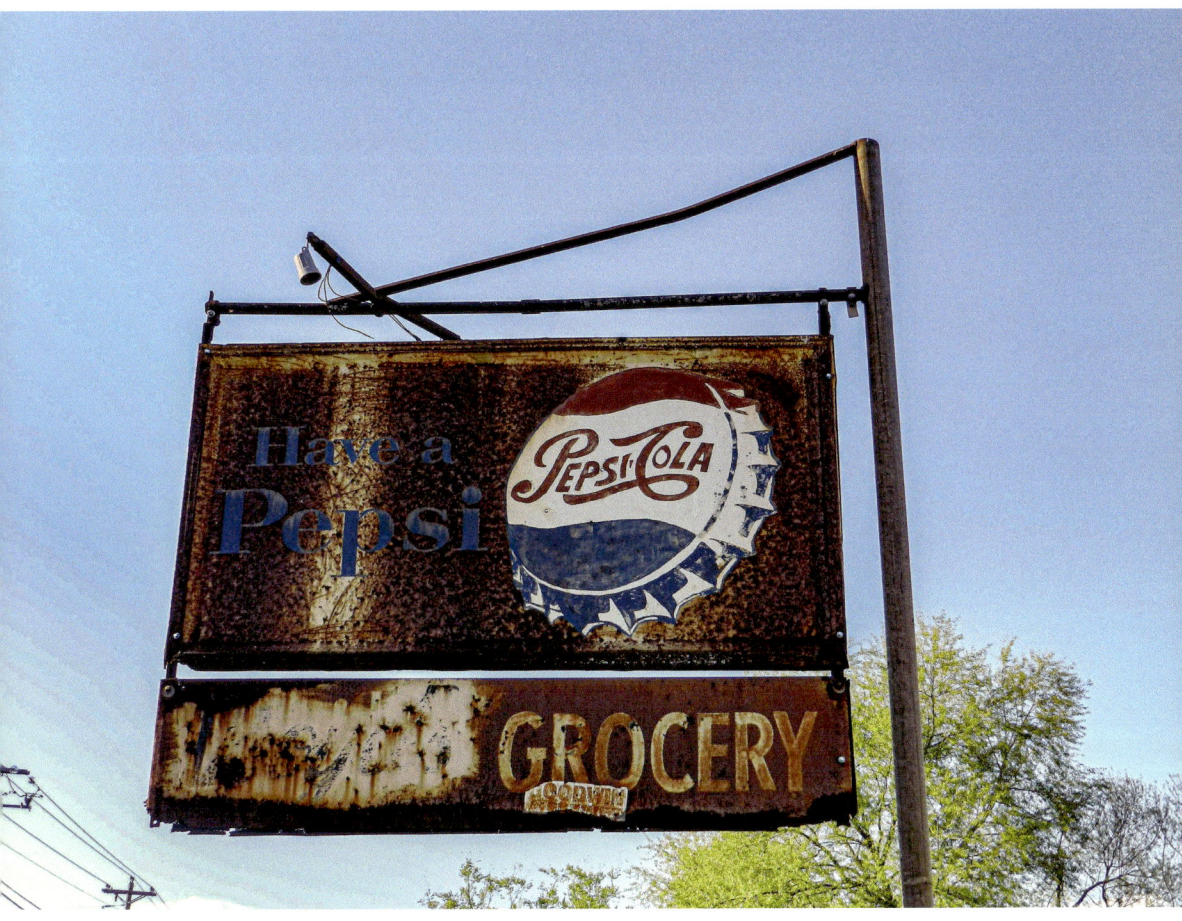

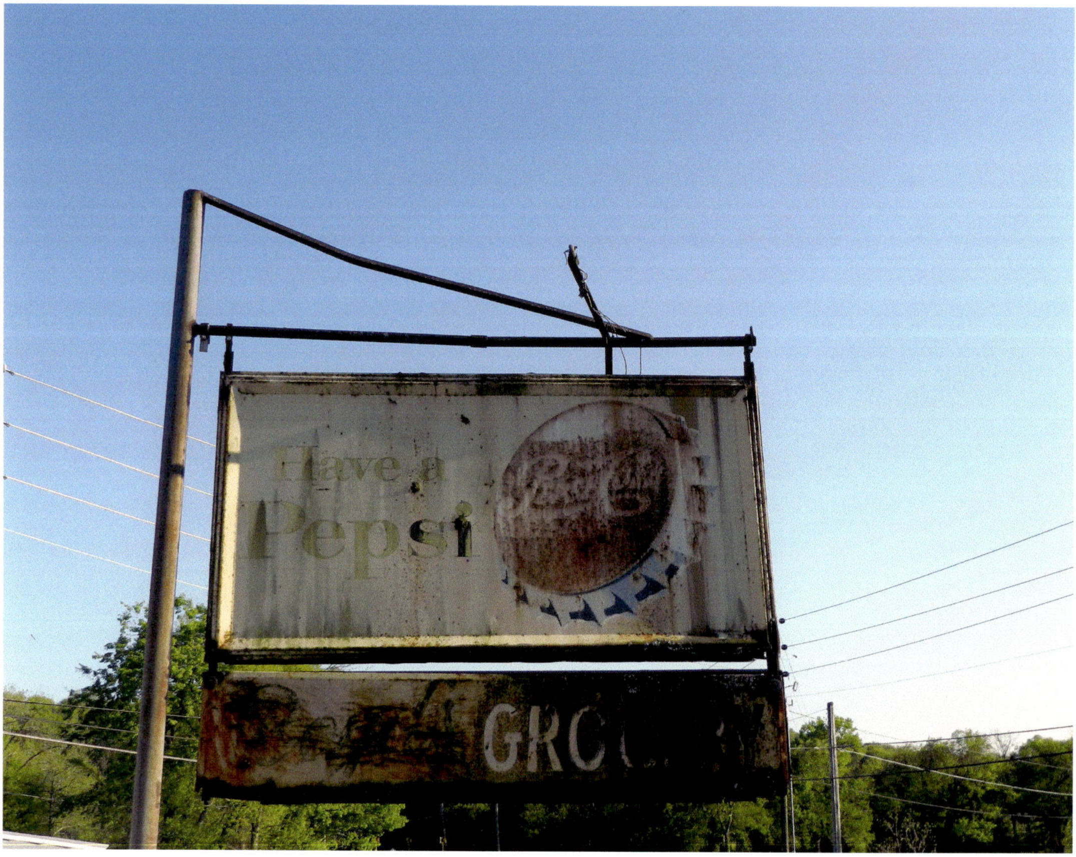
Have a
Pepsi

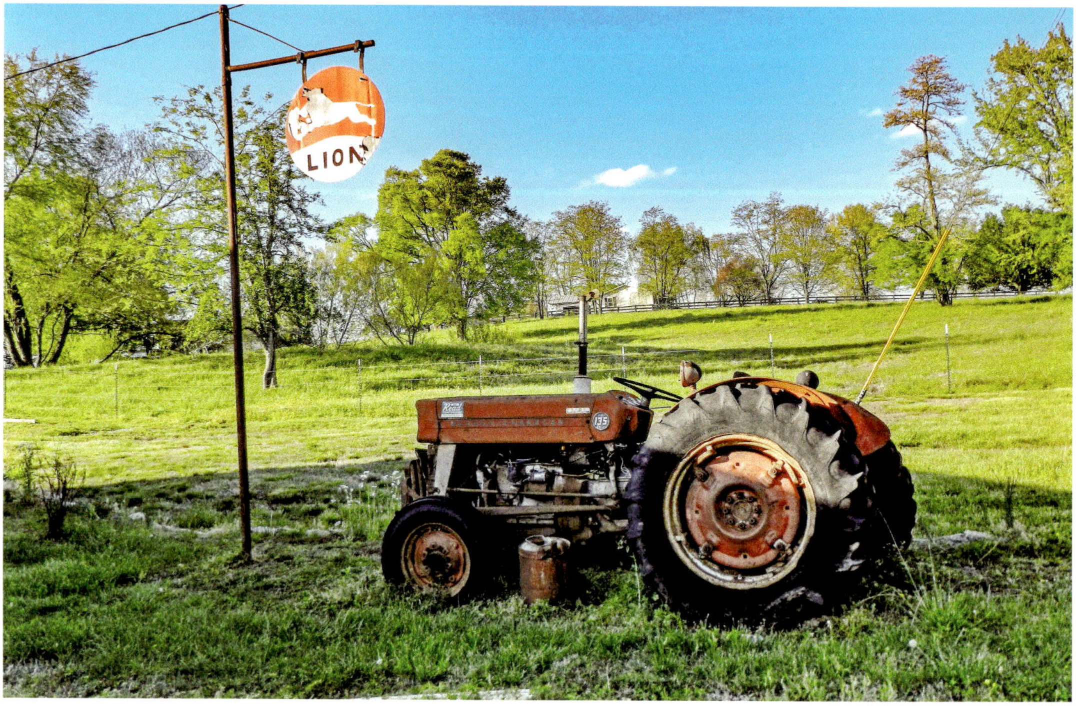
LION

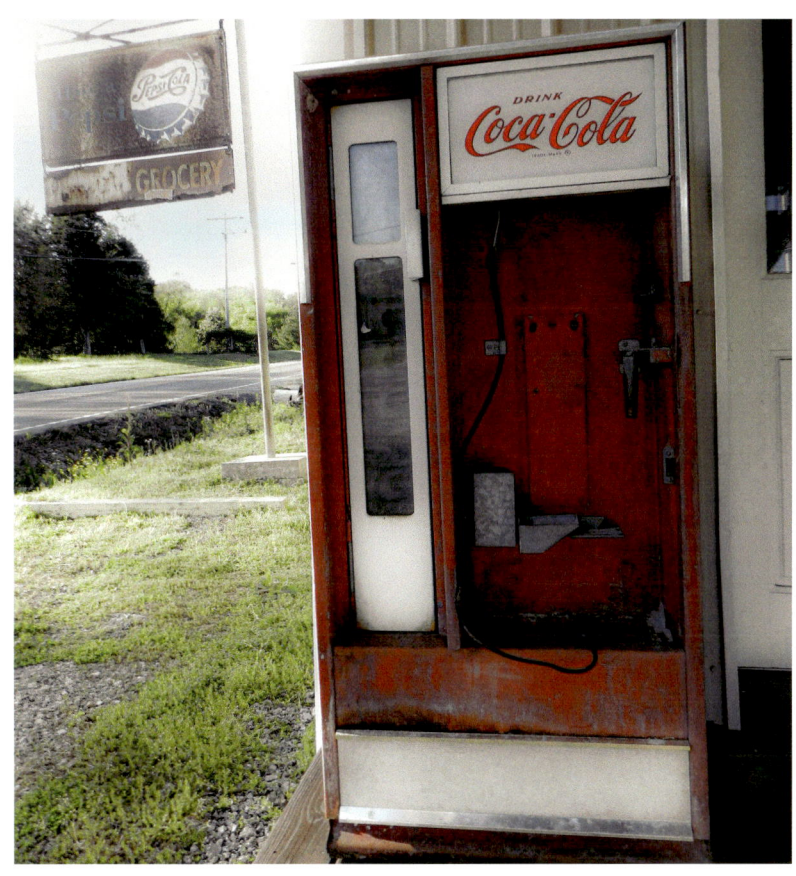

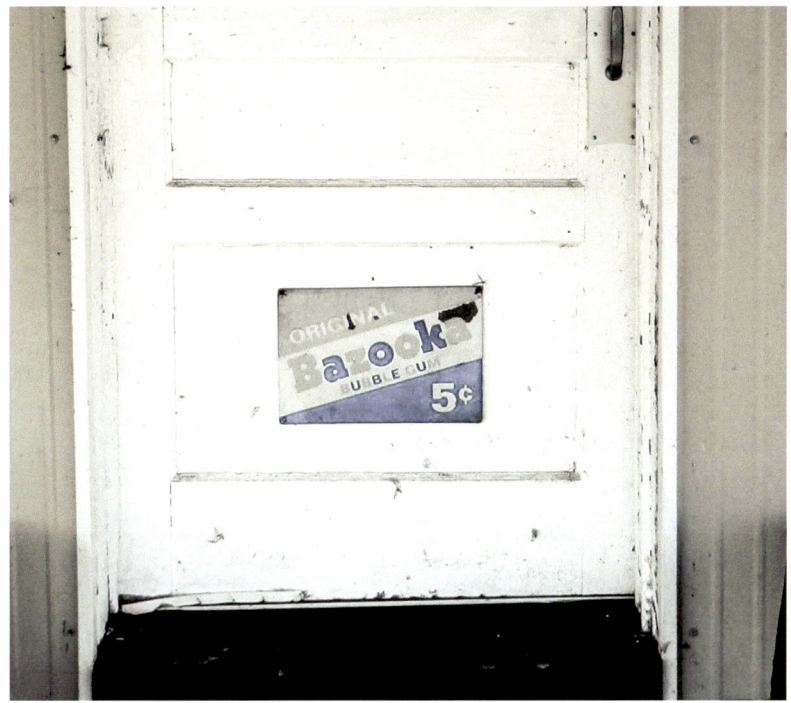

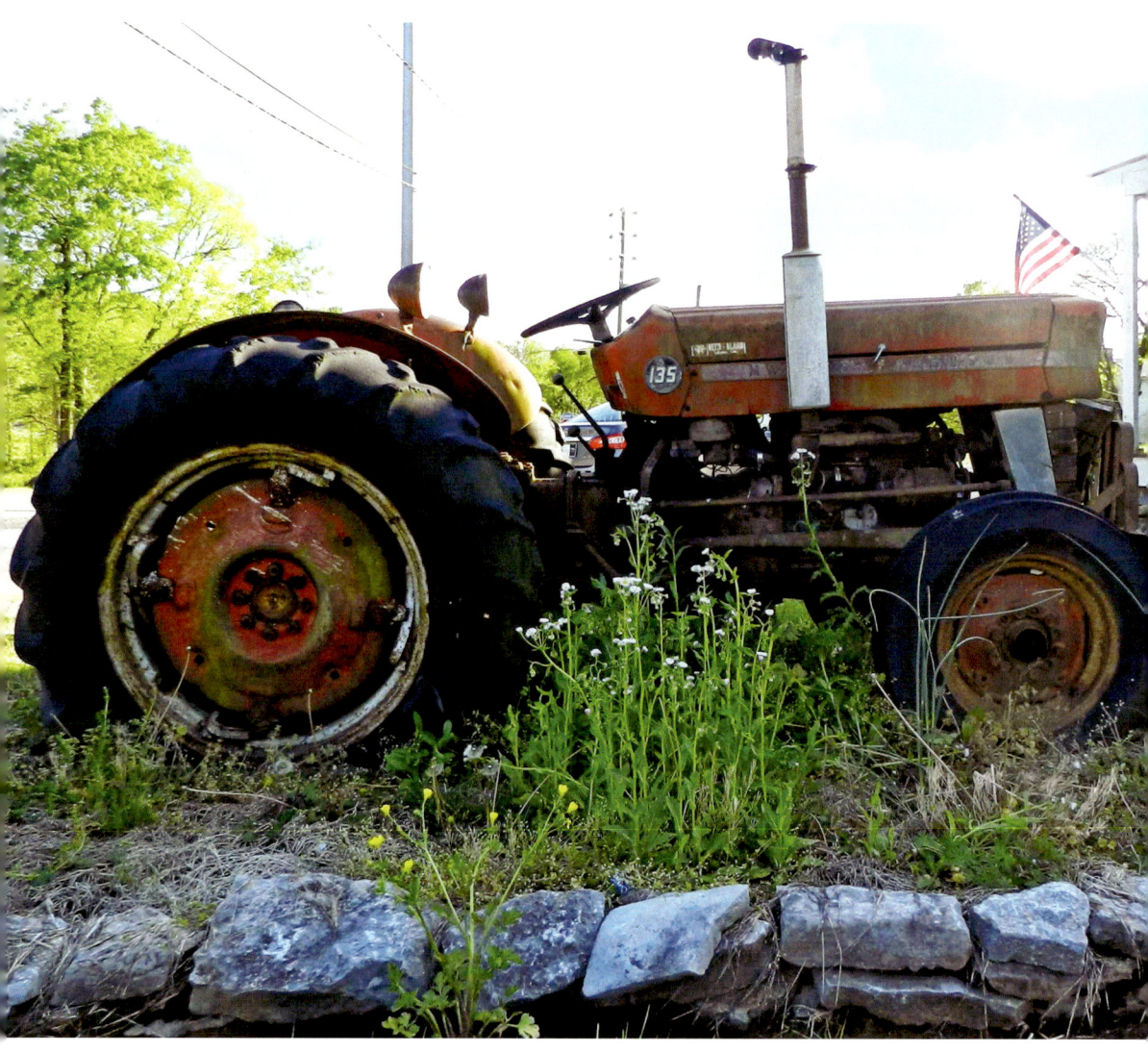

Rust Whispers Tales of Pearcy's, the Town's General Store

These photos capture a glimpse at Pearcy's General Stores past. While the pumps may be dry and the shelves less stocked, these photos capture the heart of the original location, a hub for the community for over fifty years. Though they have moved just about 100 yards down the road, they still are a small town's main spot for farm and feed supplies, the spirit of Pearcy's in Lascassas lives on! Go visit them at their new location to continue the tradition of supporting local, family-owned businesses.

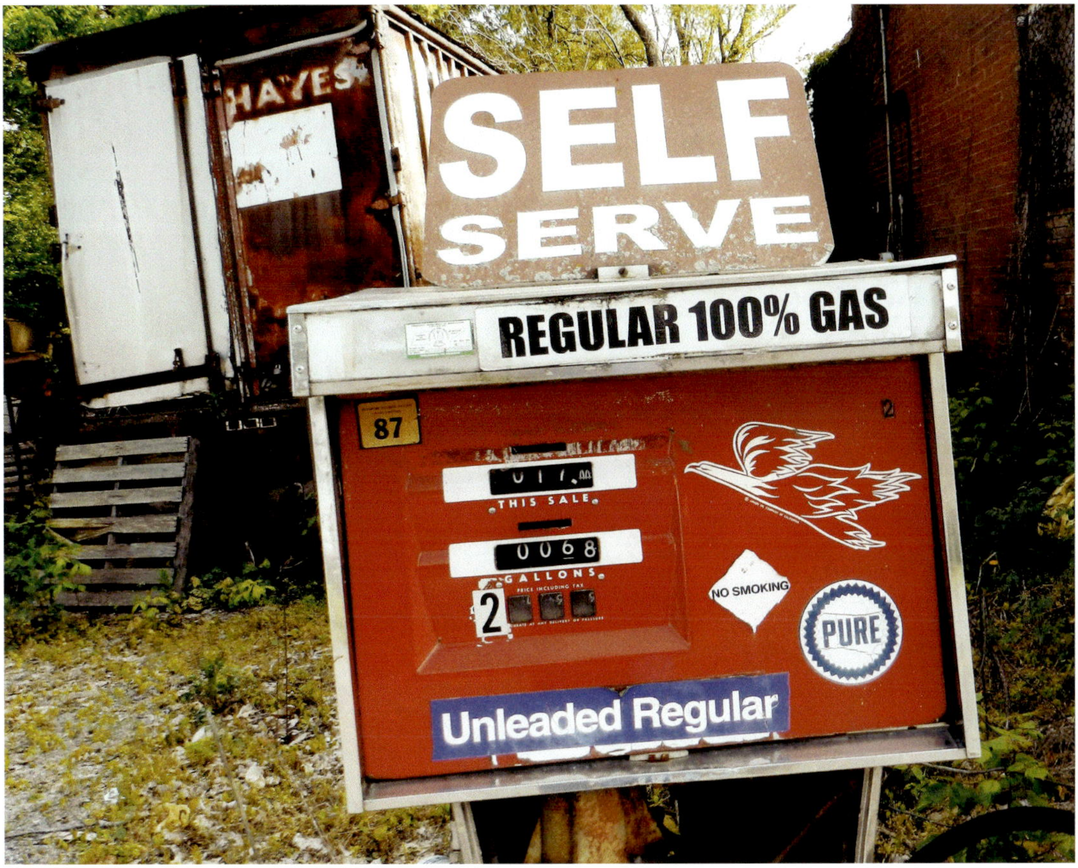

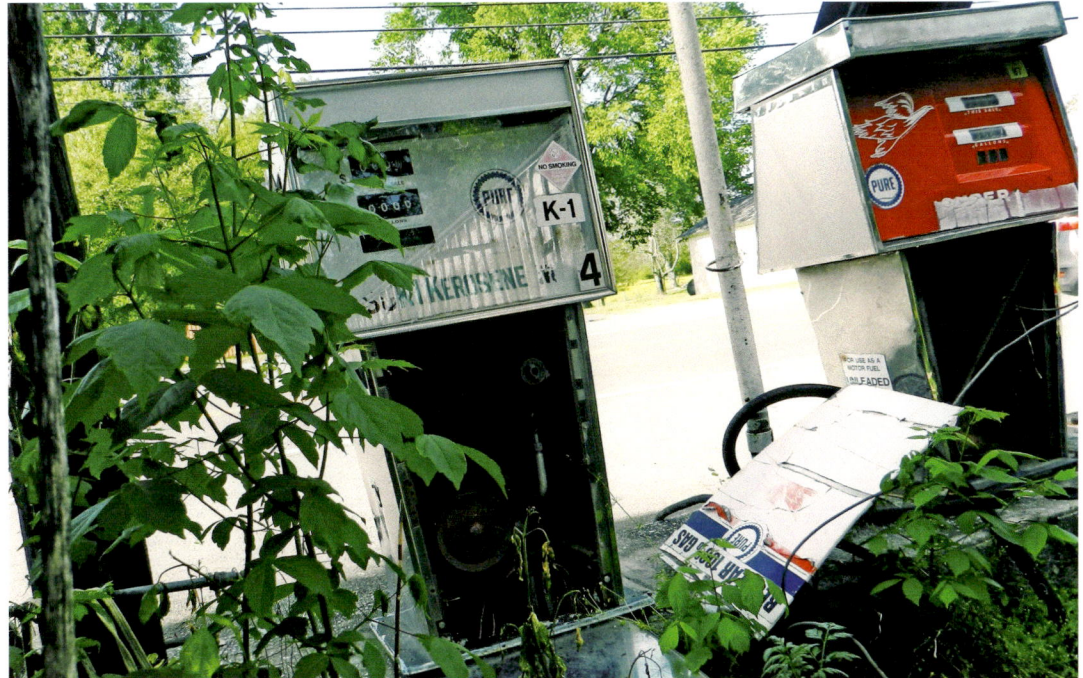

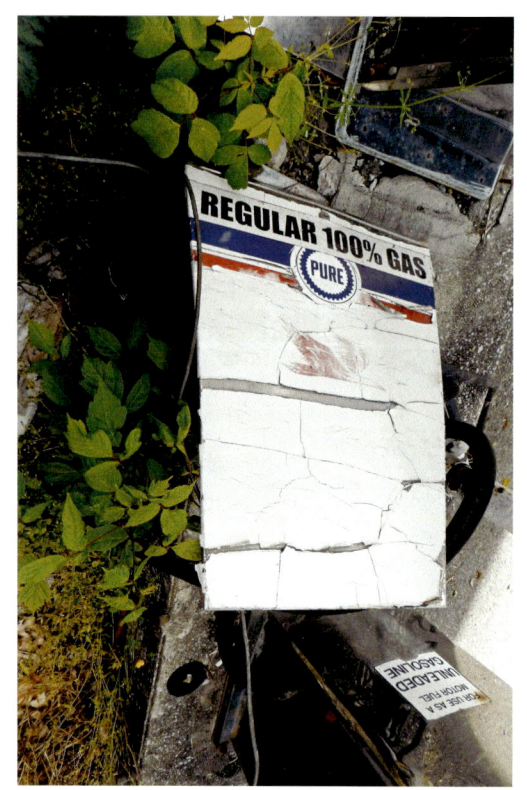

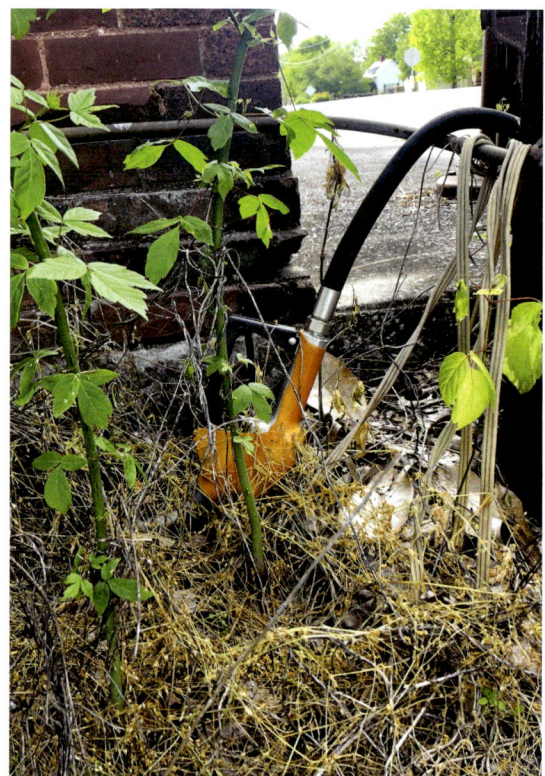

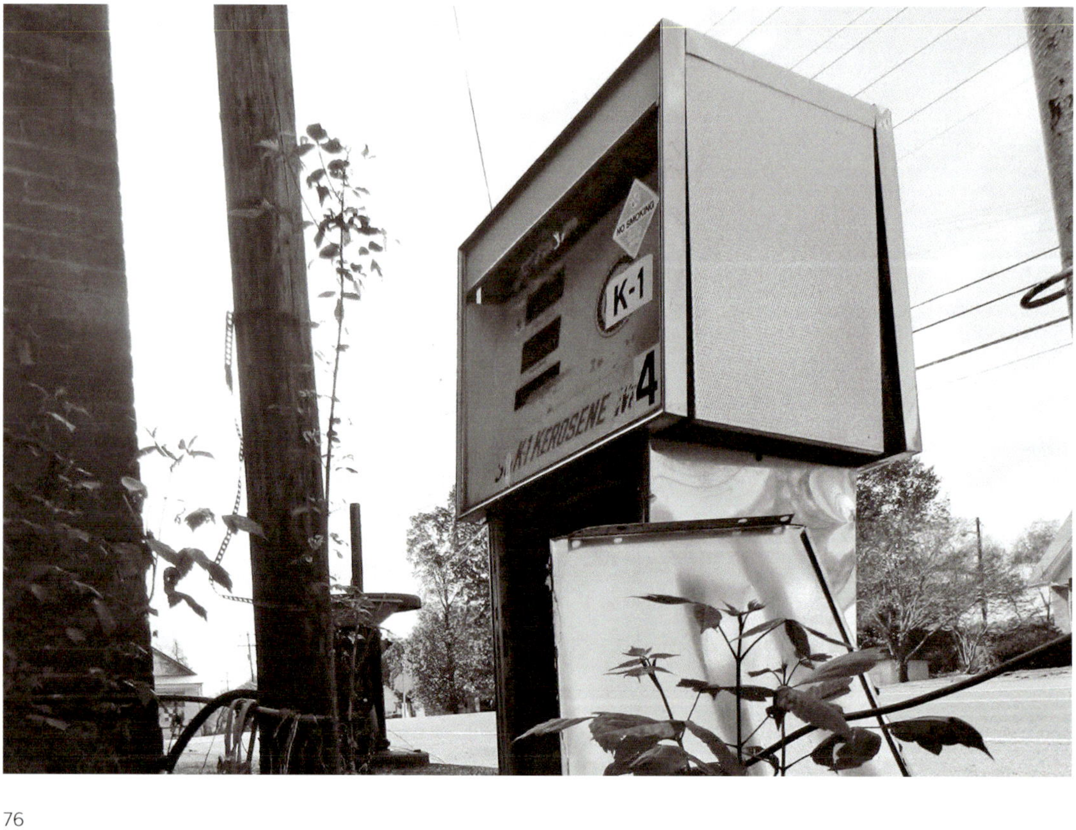

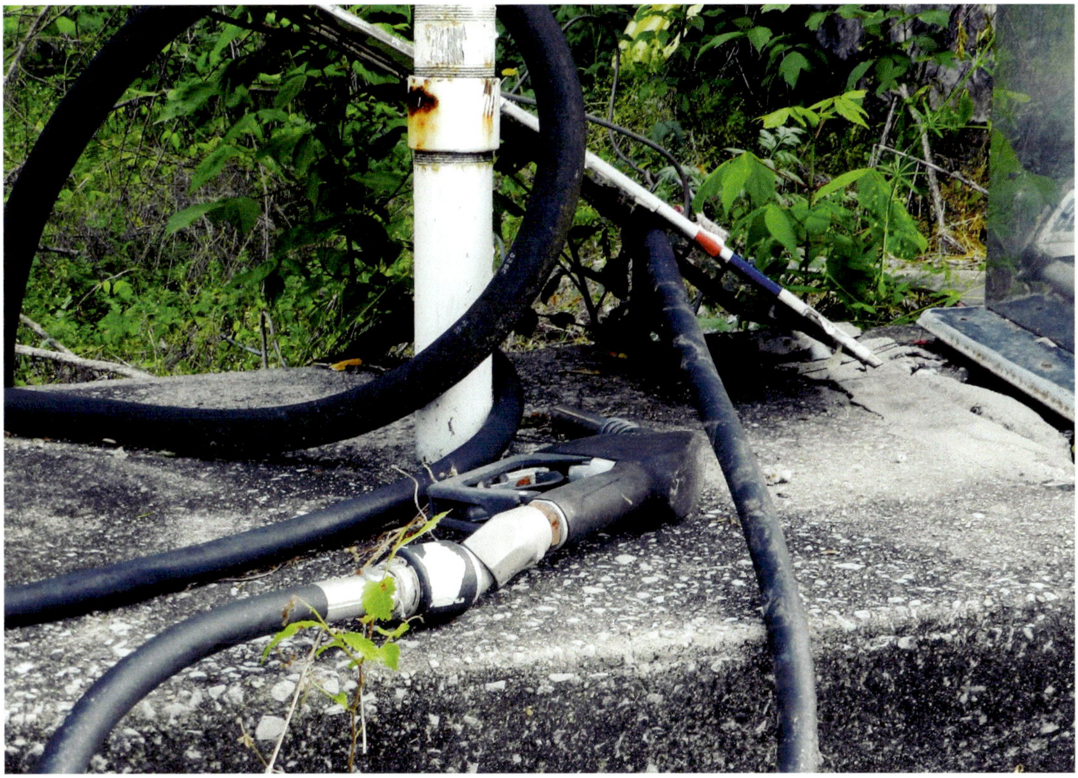

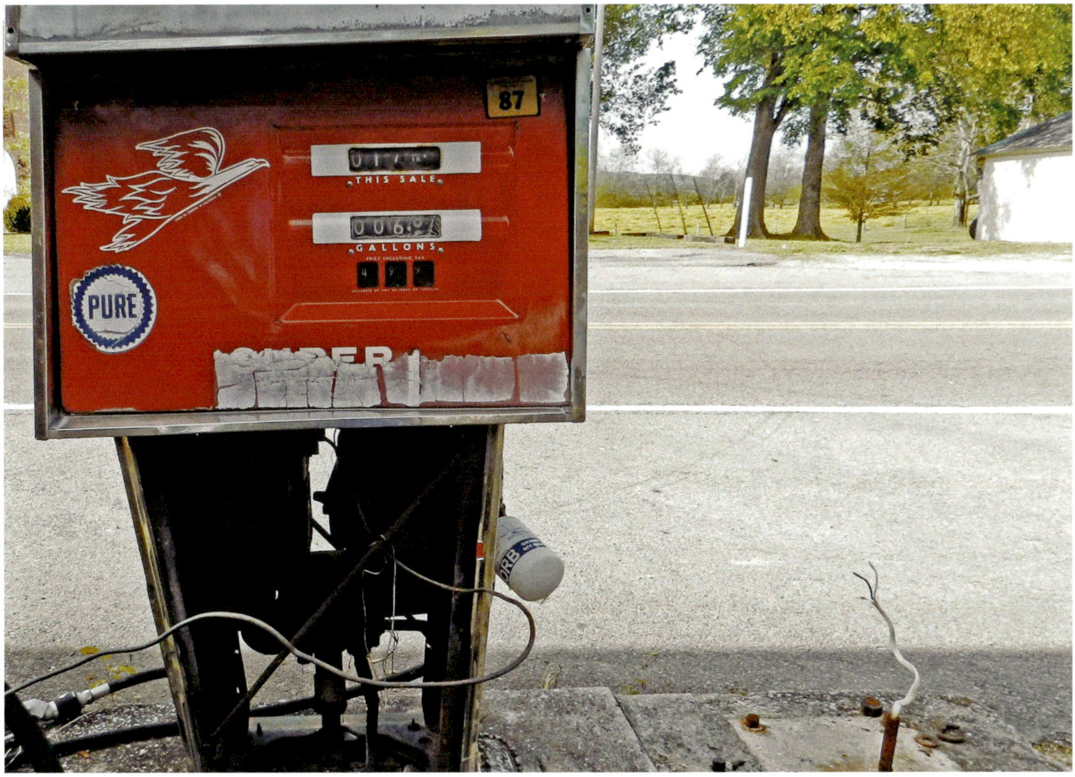
87
THIS SALE
GALLONS
PURE

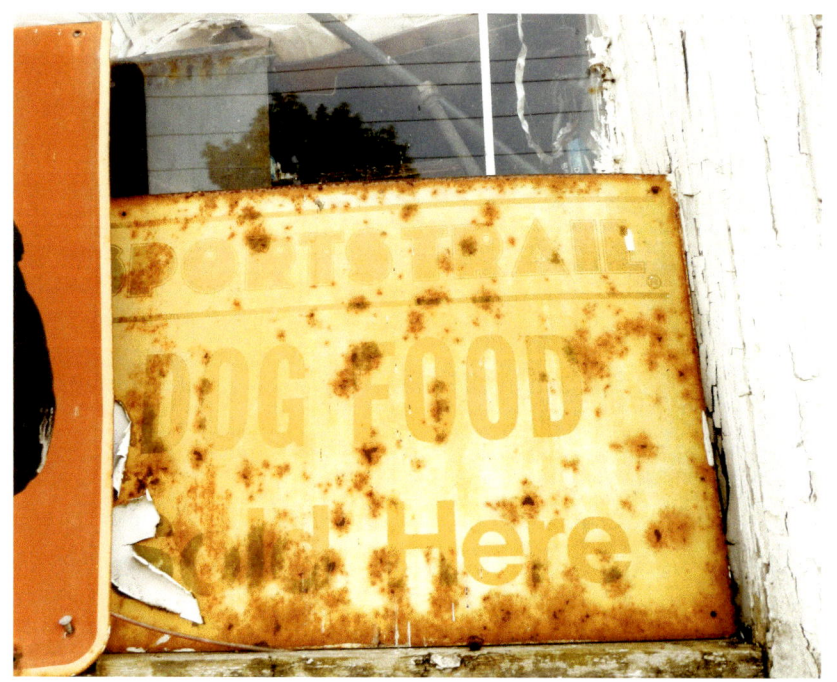

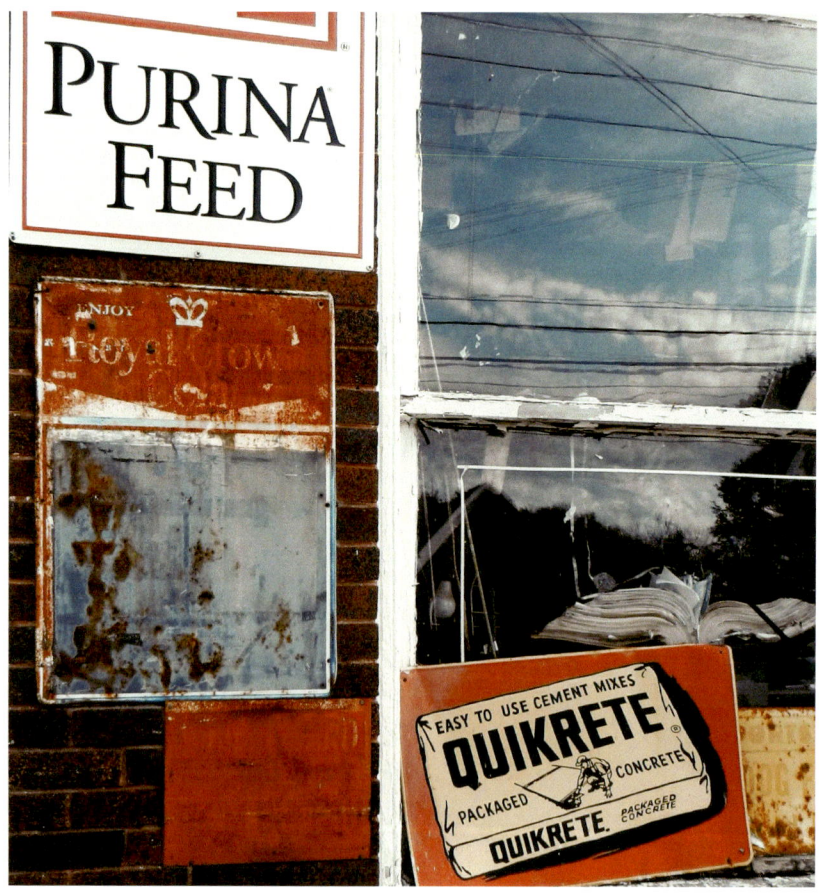

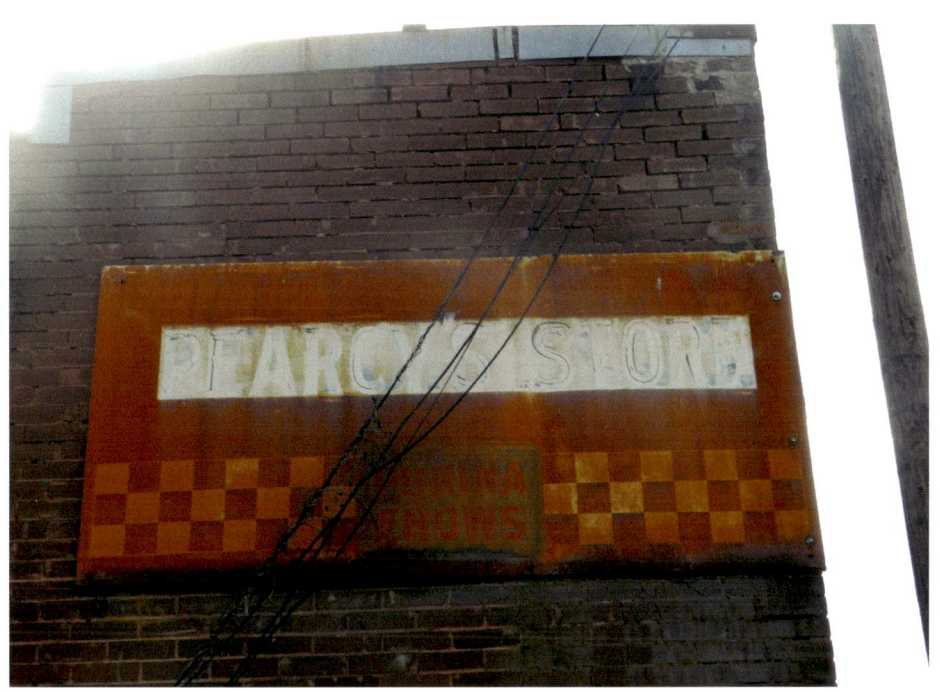

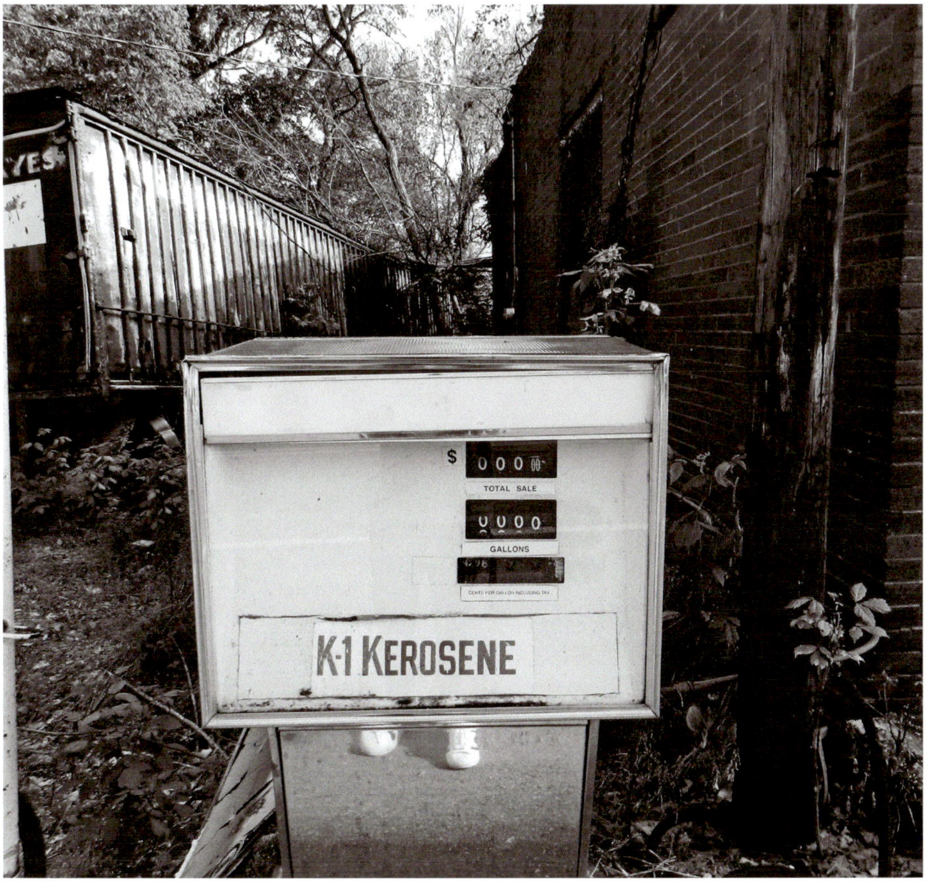

Smoke Signals from a Forgotten Past

All alone in the field stands this silent sentinel, a weathered brick chimney reaching towards the sky. It is the sole survivor of a homestead with a history as rich as the soil it stands upon. Whispers of the past linger in the air, stories of lives lived and milestones reached within these very walls. While the exact details escape me, this chimney undoubtedly holds a place in the town's narrative. Perhaps it housed a family who played a pivotal role in the community's development, or maybe it witnessed a historical event that shaped the region. One thing's for certain, this solitary stack stands as a testament to the passage of time and the enduring spirit of the past.

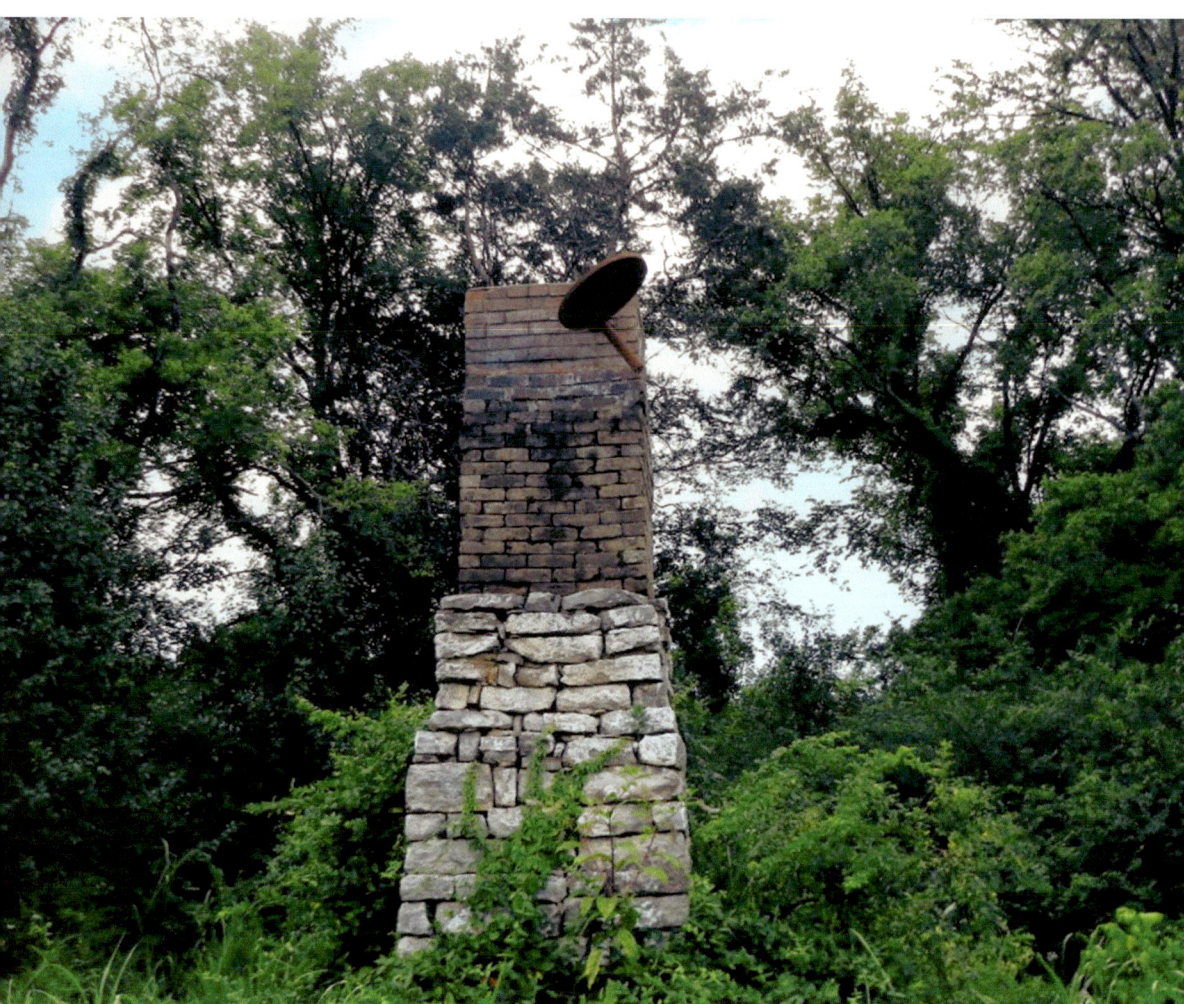

Shelved Dreams: A General Store's Forgotten Past

Carlton's General Store, a relic from a forgotten era. The slogan "We Sell Most Anything" still lingers faintly on the peeling paint, a ghost of its former promise. Green algae paints the walls an eerie hue, a stark contrast to the vibrant goods it once stocked. A lone cash register, frozen in time, displays a cryptic message, hinting at the store's final moments. In the corner, barrels from barrel racing lie stacked, a testament to simpler times. Carlton's may be abandoned, but these photos capture the essence of a community hub, leaving you to wonder about the stories whispered within its dusty shelves.

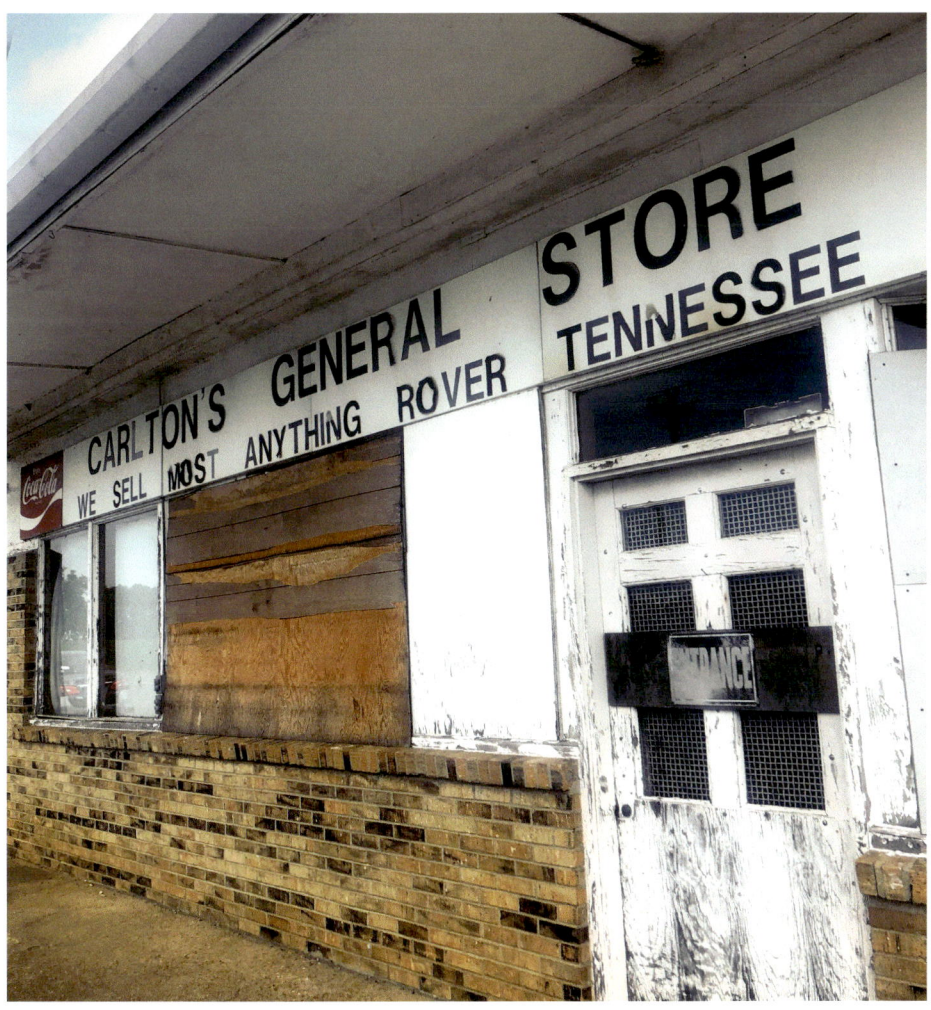

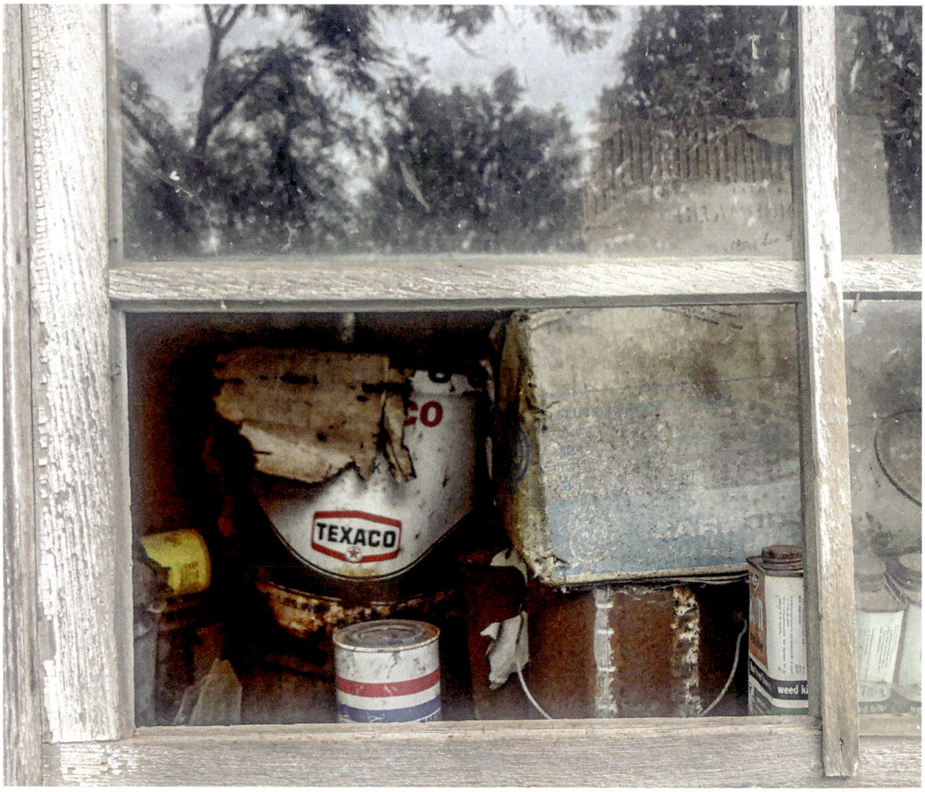

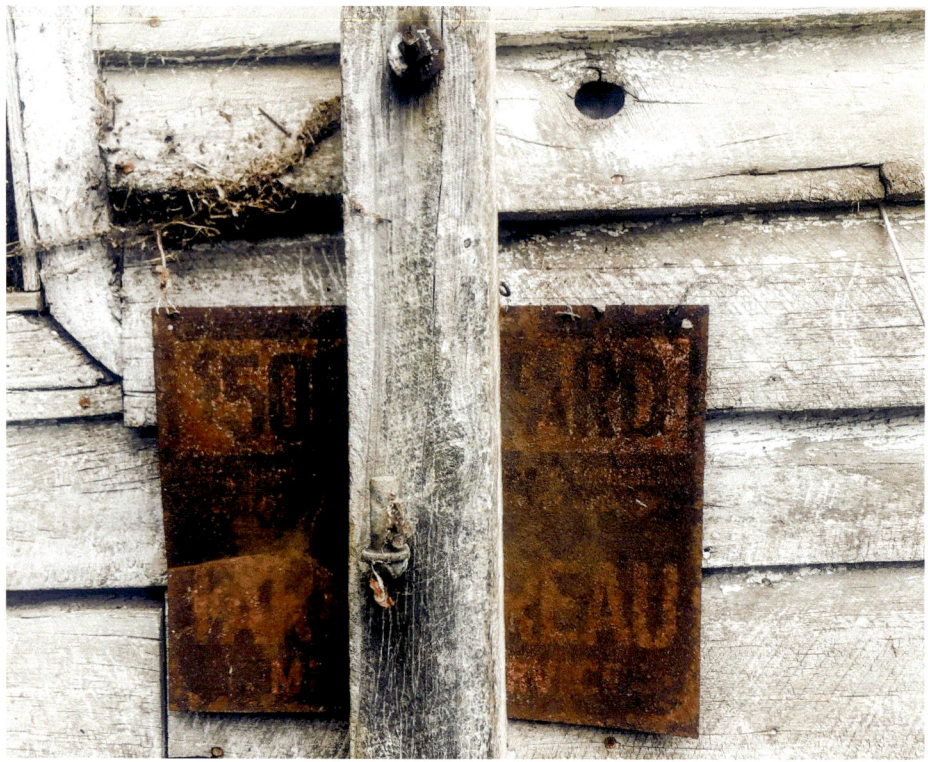

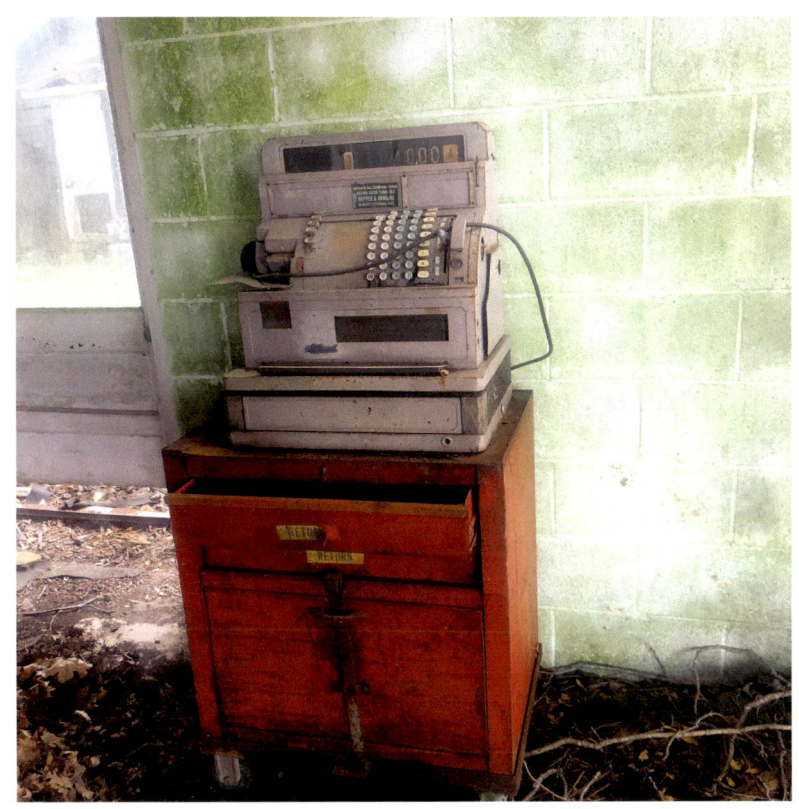

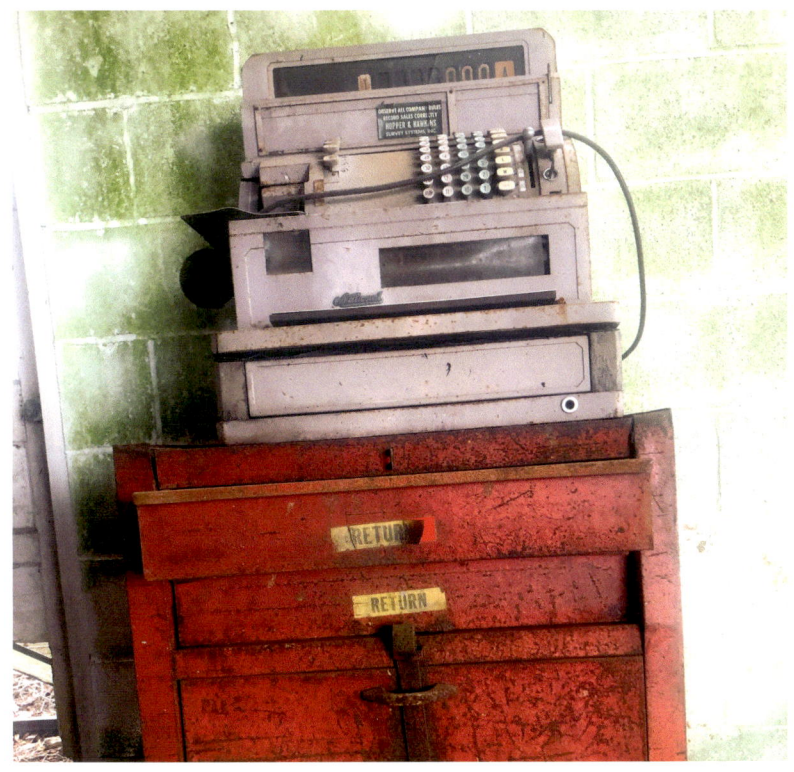

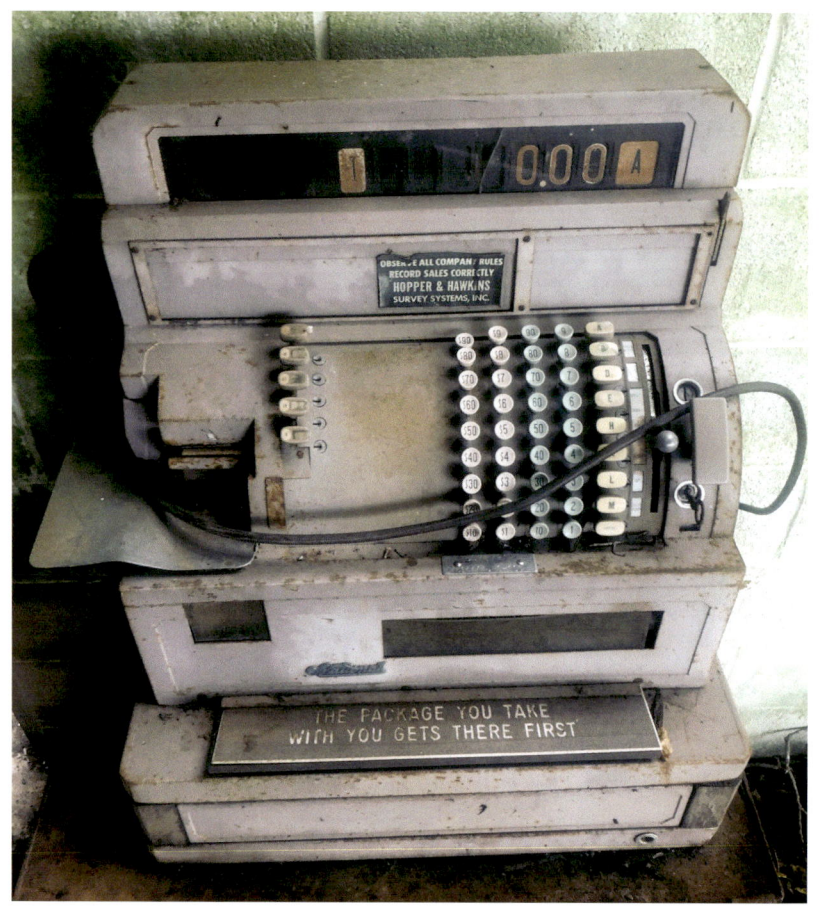

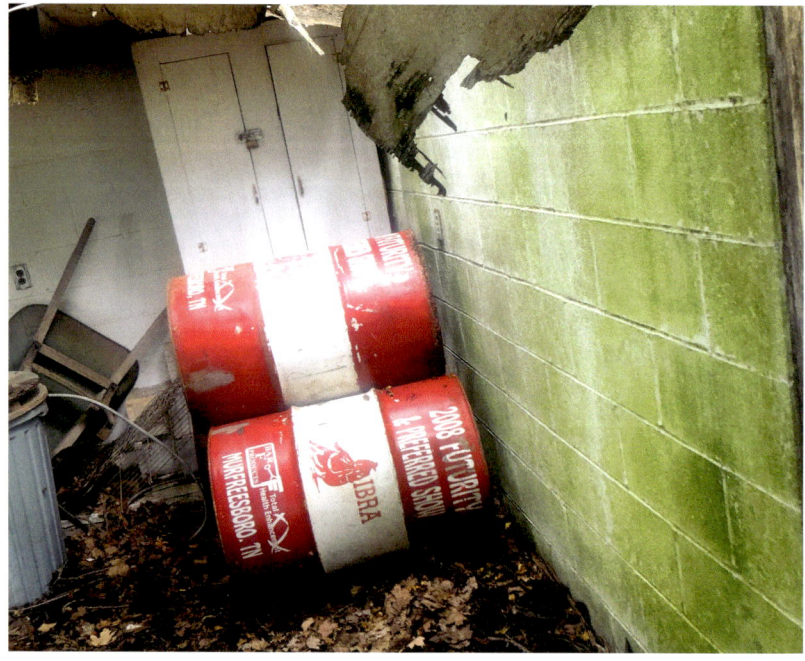

Free Fireworks, Empty Promises: An Abandoned Enigma

This weathered building whispers a forgotten promise but caught my eye from the roadway. A rusty desk stands guard by a crude "Fireworks for Free" sign. Did a celebration fizzle out? The truth, like the fireworks, has vanished, leaving a mystery: what went boom?

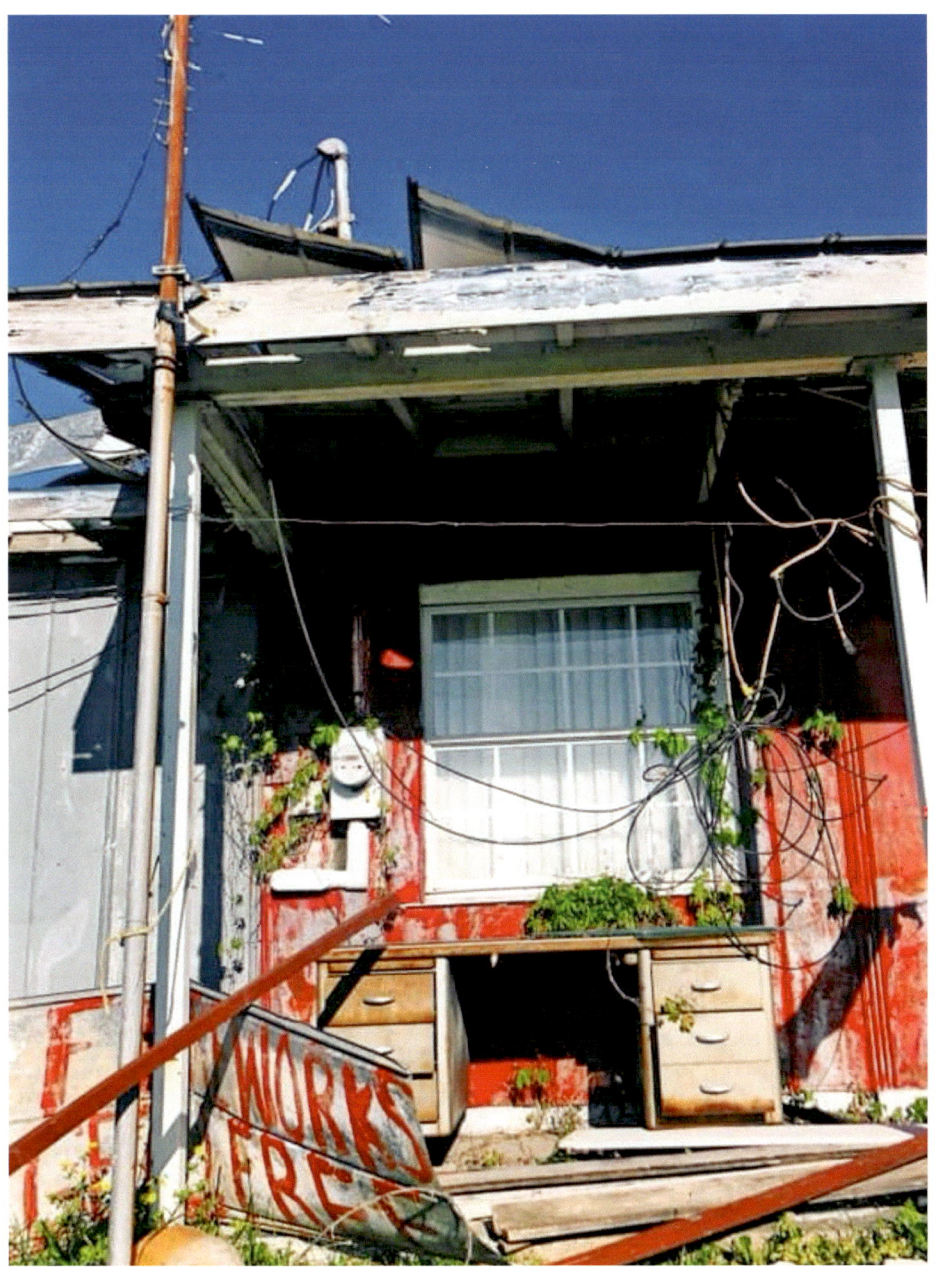

From Storefront to Sanctuary:
The Untold Story of a Condemned Care Center

This building, once a bustling store, transformed into a haven for veterans, offering solace and support for countless years. Laughter and camaraderie likely filled these halls, replaced now by an unsettling quiet. A condemned sign stands guard, a stark reminder of the hazardous conditions lurking within. Faded paint and empty rooms whisper stories of lives lived and sacrifices made. While a new chapter awaits this space, these photos preserve a piece of history, a testament to the unwavering dedication to those who served.

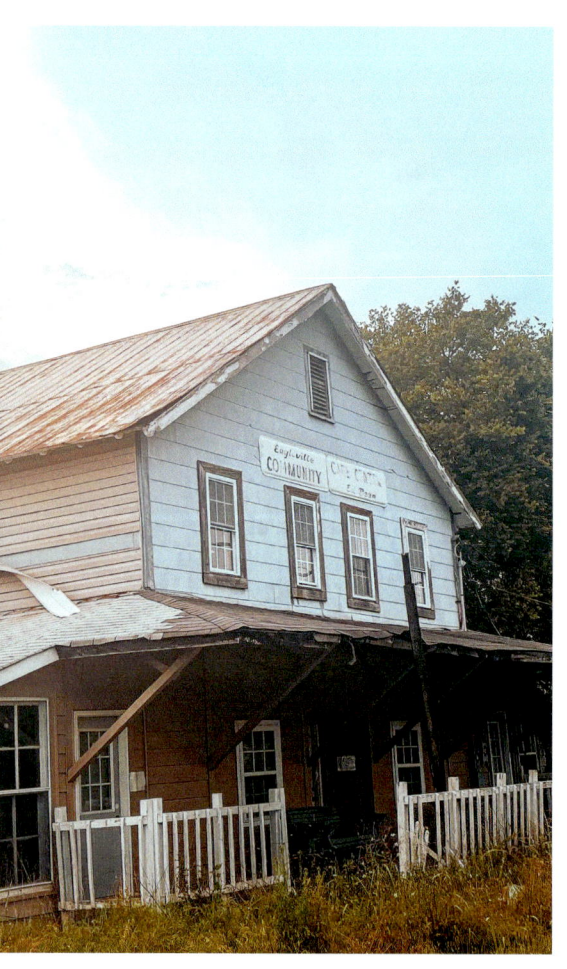

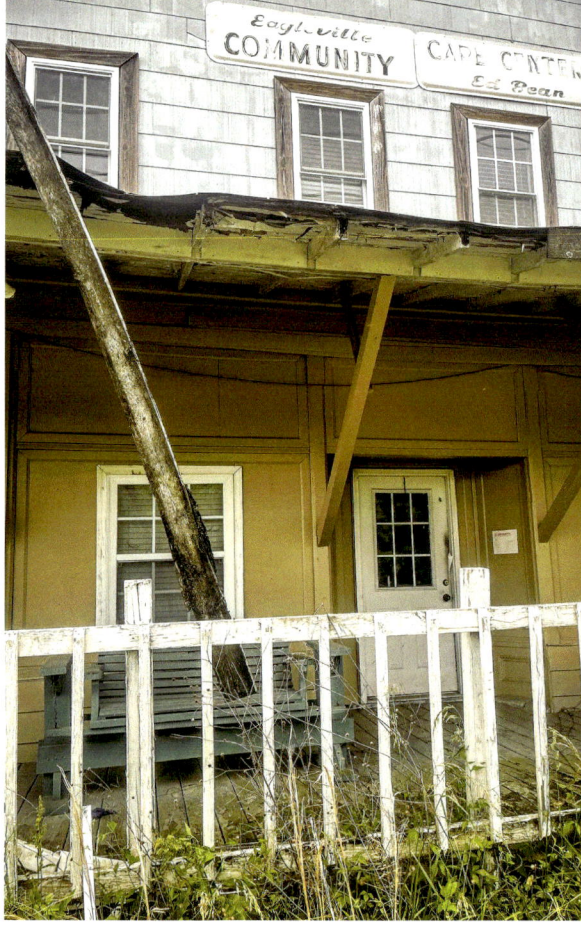

Out of Stock, Full of Memories: A Faded '80s Grocery Store

A once-bustling corner store in the Blackman community of Murfreesboro. Empty shelves whisper stories of forgotten staples and friendly faces. Dust motes dance in forgotten sunlight filtering through grimy windows. A faded awning hints at a weathered brand sign, a silent promise of daily necessities. Though the aisles are silent now, these photos capture a snapshot of a bygone era, a place where the community came together over a loaf of bread and a friendly chat.

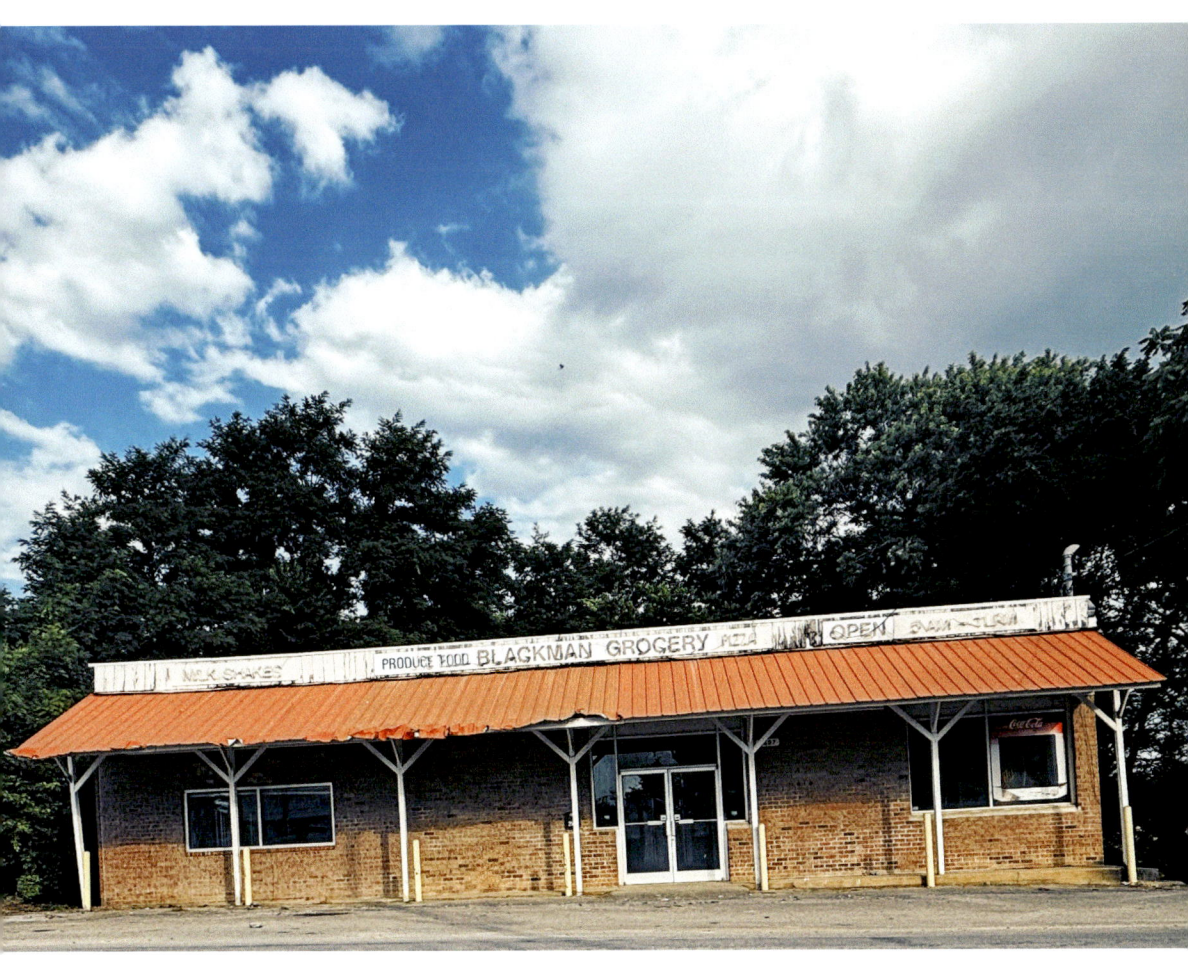

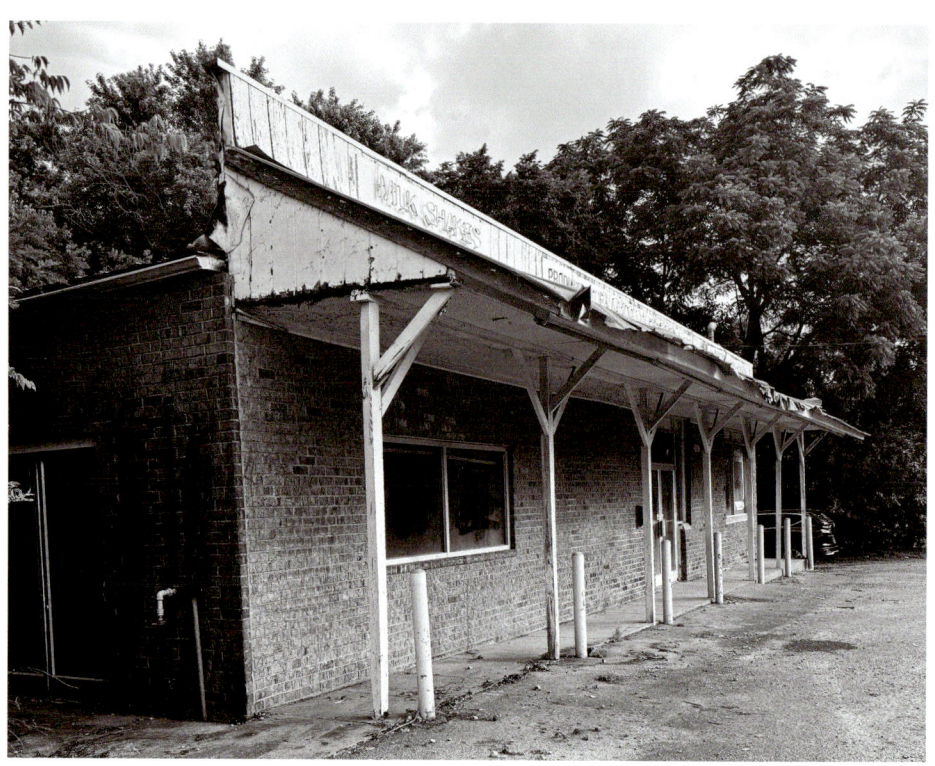

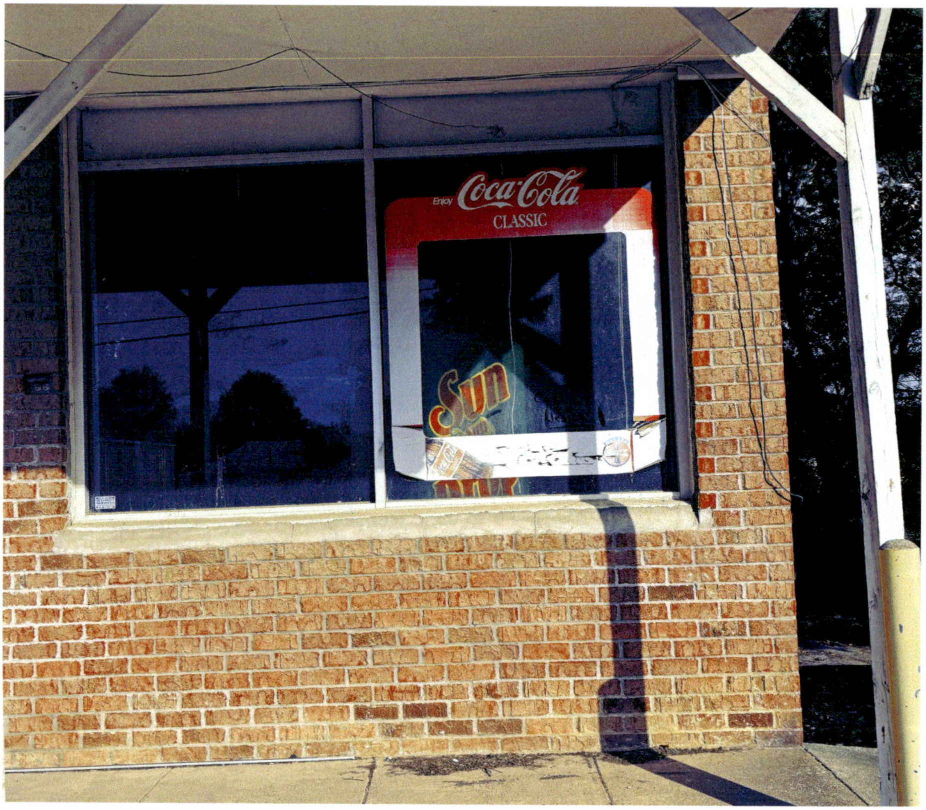

Abandoned Americana: The Faded Charm of Loafers Gap Grocery

Weathered gas pumps stand guard, their chrome gleaming with a faint echo of their former glory. Above them, a chorus of faded steel sings a silent brand loyalty anthem: a vibrant red Pepsi sign, a classic script Coke logo, and whispers of other forgotten favorites. Antique glass Coke bottles, their contents long evaporated, stand beside two rocking chairs, swaying gently in the phantom breeze of memory. This is a place where time has pressed pause, leaving behind a collection of relics that whisper tales of simpler times, of friendly chats over gas fill-ups, and the sweet satisfaction of an ice-cold soda on a hot summer day.

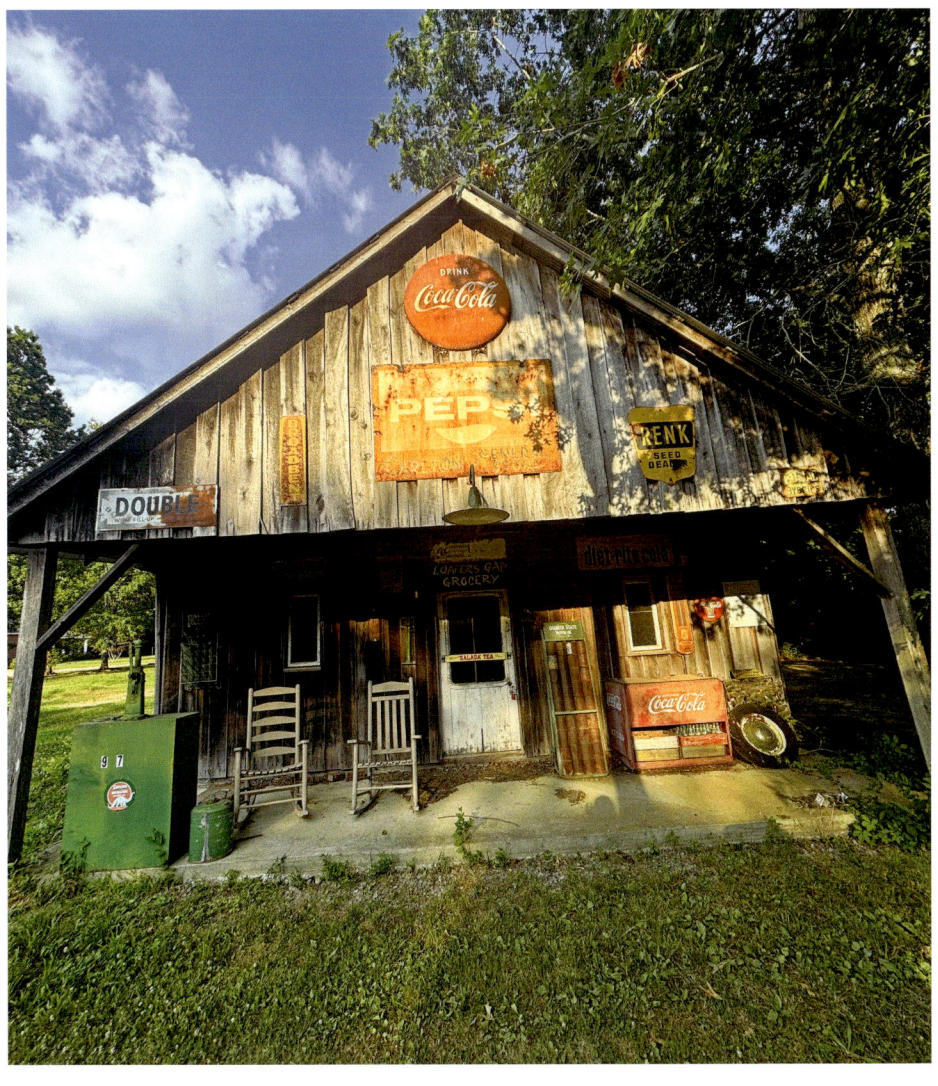

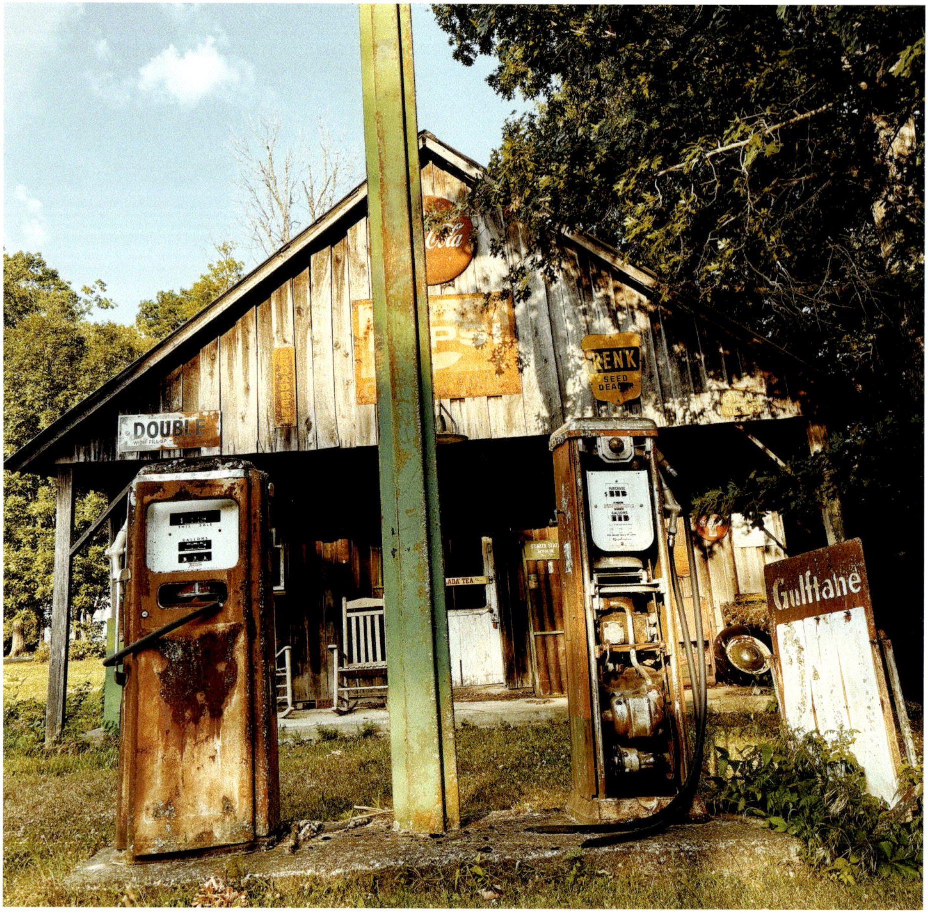

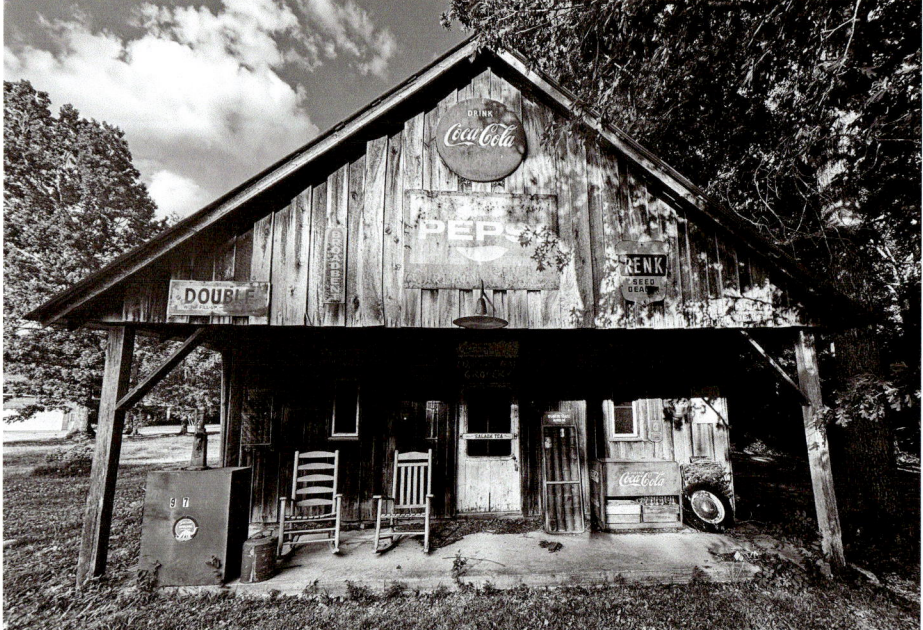

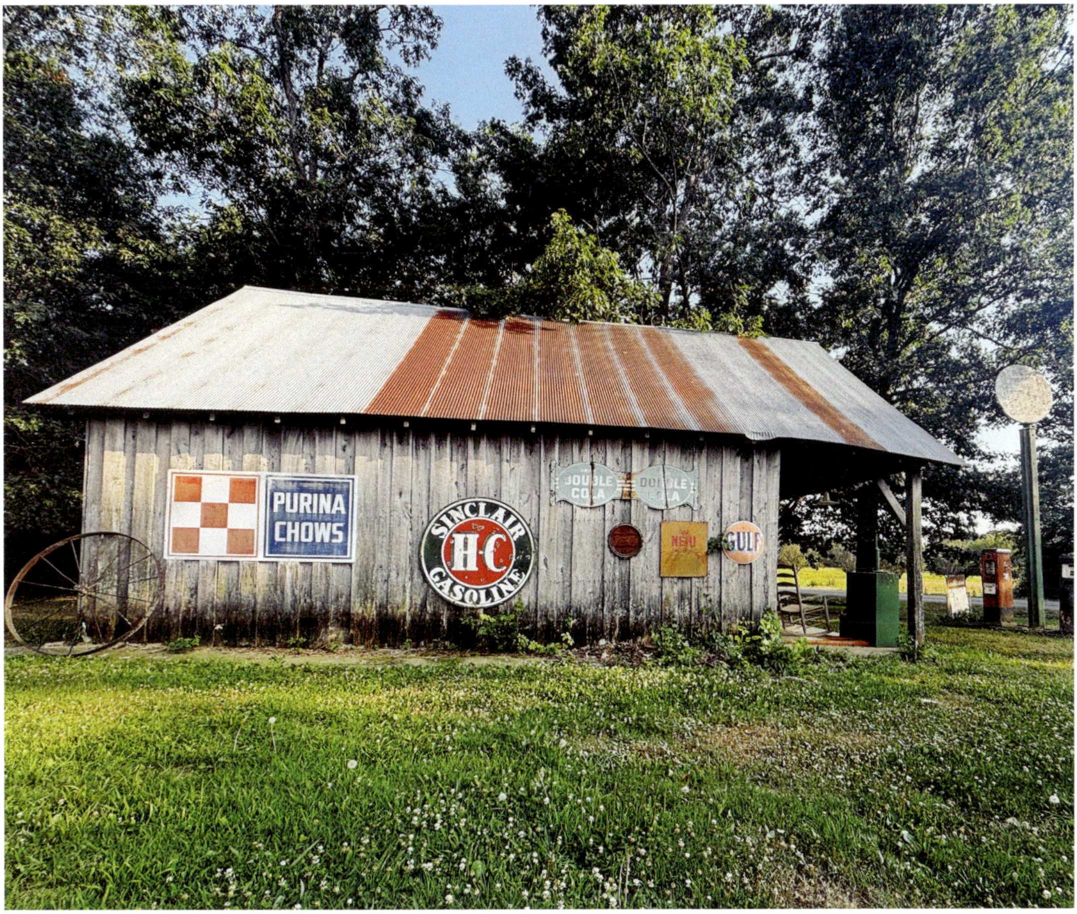

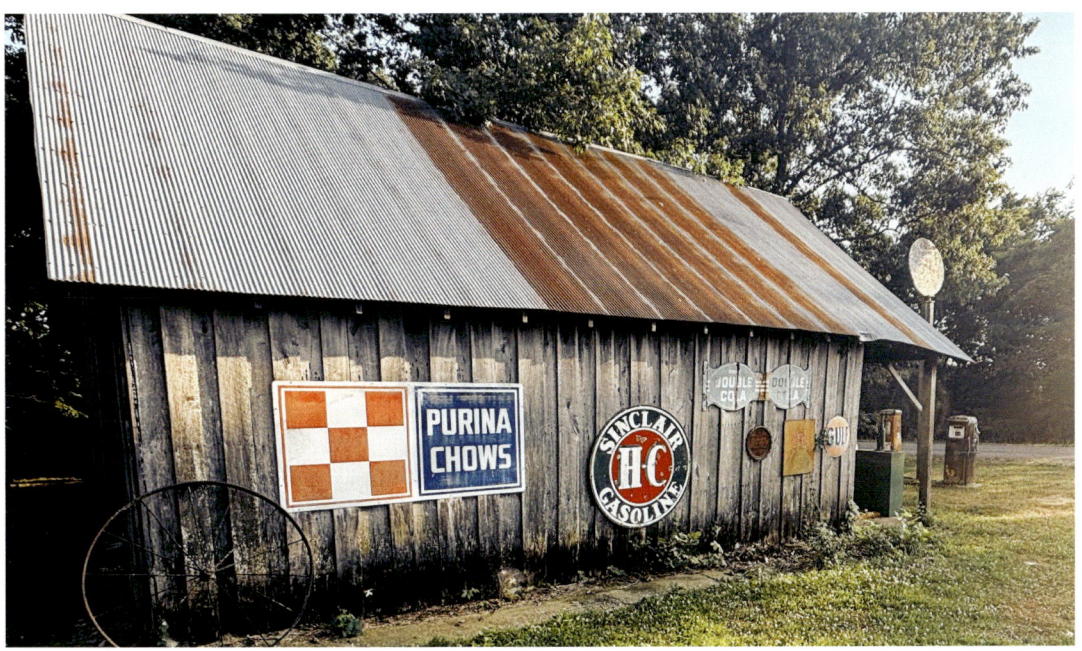

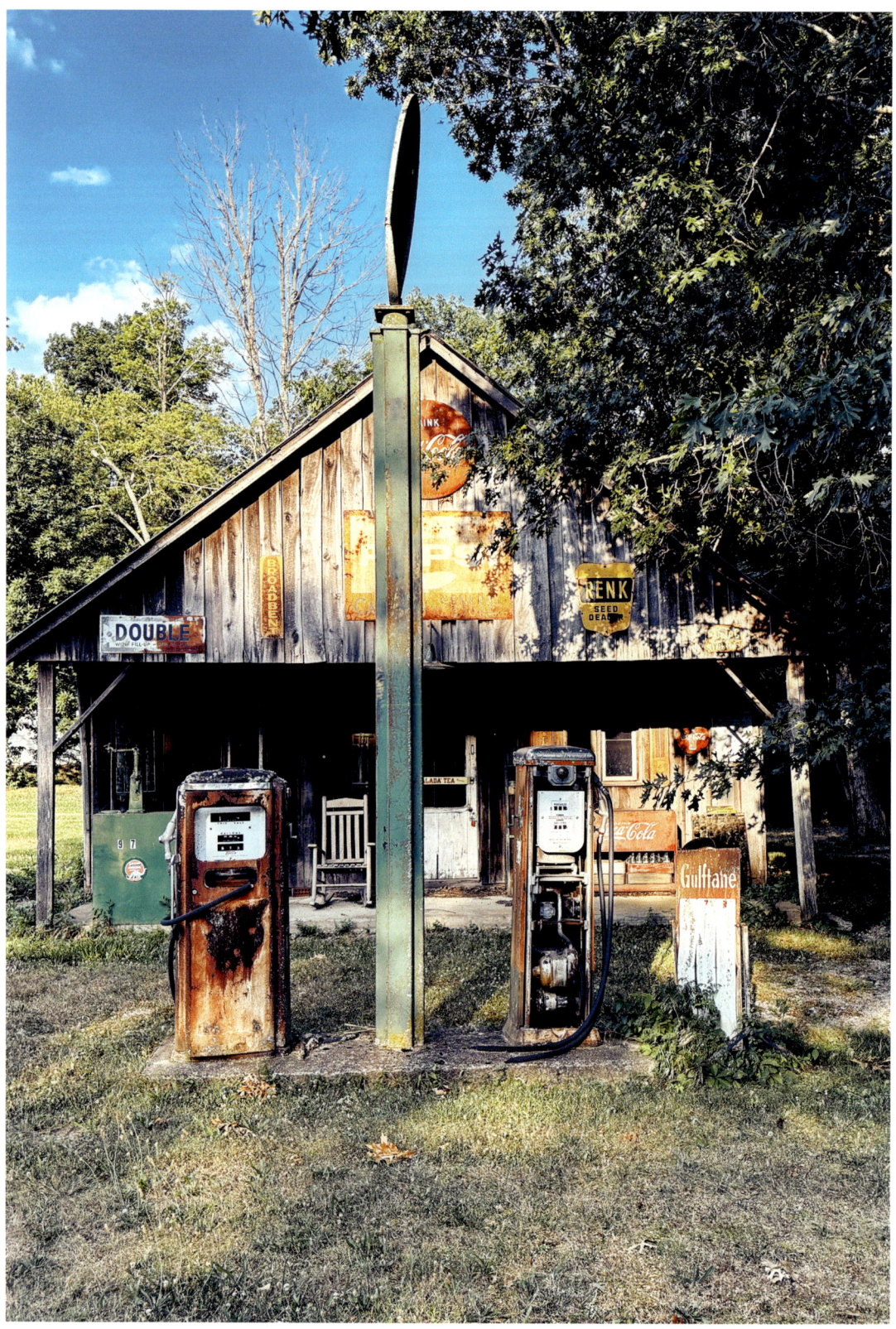

Unclaimed Hopes: A Rusted Reward and a Barn's Silent Plea

Nestled among a tapestry of emerald vines, a weathered barn stands sentinel, its secrets cloaked by nature's embrace. Sunlight peeks through gaps in the aged wood, casting an ethereal glow on the peeling paint and a solitary reward sign clinging to the door, standing as a faded testament to a lingering mystery.

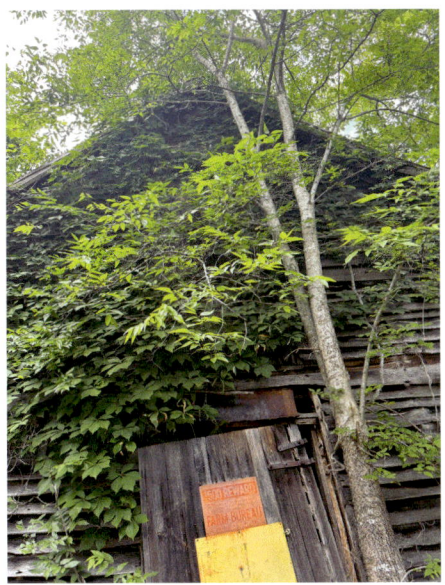

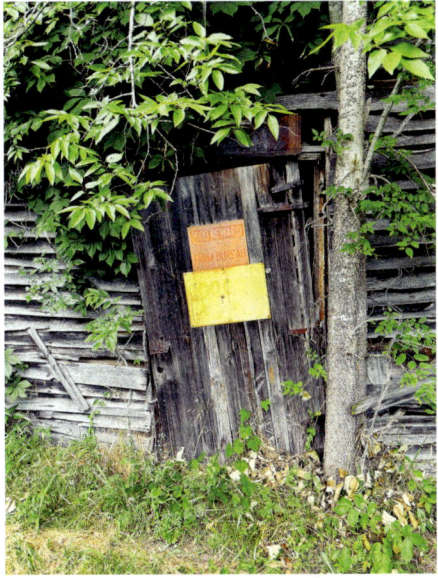

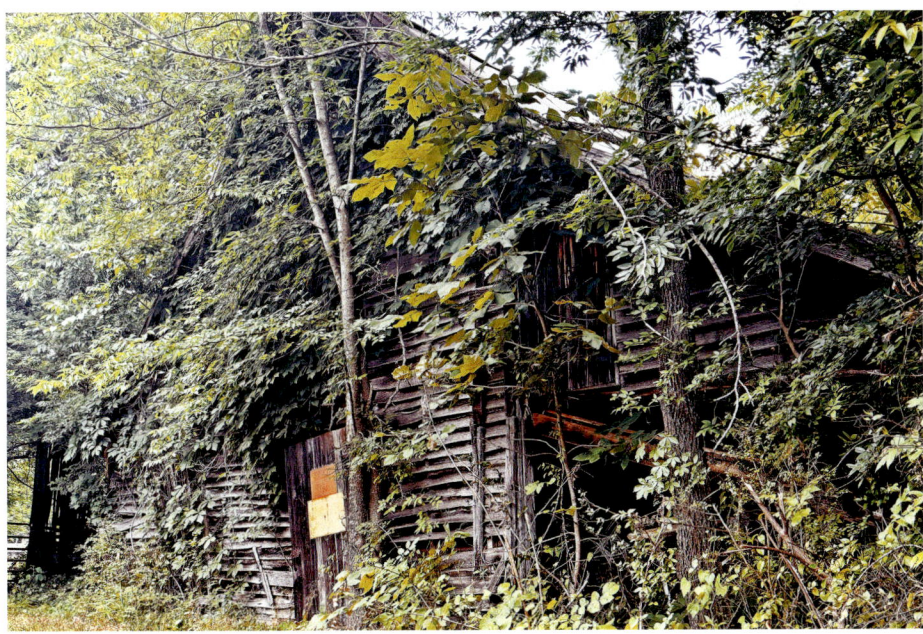

FINAL NOTE

My fascination with abandoned places began as a teenager cruising the backroads of my quiet town. These crumbling structures, peeking through trees or nestled beside forgotten paths, sparked my curiosity. What stories did they hold? Soon, exploring and photographing them became my passion, a way to escape the ordinary and delve into the extraordinary.

Over time, my motivation evolved. It was not just about the thrill of discovery; it became a mission to preserve the whispers of the past. The rooftops visible through the trees, remnants of houses along bustling roads, once sheltered families whose lives remain a mystery. The rusting vehicles in fields were not just forgotten machines; they were once family wagons carrying dreams on road trips, or a farmer's loyal companion, putting food on our tables. The empty shelves and faded signs of corner stores tell a different tale. These were oases for weary travelers, community hubs where neighbors exchanged greetings and stories.

For over two centuries, Murfreesboro and its surrounding areas have witnessed a remarkable evolution. They were part of a pivotal Civil War battle, etched in history as the bloodiest in terms of casualties. Post-war, the community thrived on agriculture, with dairy and cattle farms feeding the region. By 1853, Murfreesboro earned the nickname "Athens of Tennessee" due to its blossoming educational institutions. Despite the war's devastation, Murfreesboro's spirit of resilience shone through, allowing it to rebuild while other areas remained crippled.

War, agriculture, and education are the cornerstones upon which this community was built. Yet, these abandoned structures are equally important. They stand as silent testaments to the lives lived, the hardships overcome, and the foundation upon which Murfreesboro flourishes today. This book serves as a tribute to these forgotten spaces and the stories they hold, a reminder of the vibrant past that shapes our present and future.

Thank you for joining me on this journey through forgotten corners. I hope the photos and stories inspired you. My explorations will continue, fueled by a love of discovery and a dedication to preserving these pieces of history. Remember the golden rule of urbex photography: "Take only pictures, leave only footprints." Follow me on my future endeavors, whether in a new book or on social media. Until then, explore the world with an open mind and a curious spirit.

ABOUT THE AUTHOR

Eros Russell, a photographer and urban explorer originally from East Tennessee, found his passion amid the quiet landscapes of rural southern towns. As a teenager, with little else to occupy his time, he would navigate the winding backroads, honing his keen eye for photography. It was during these explorations of abandoned places that he discovered his profound fascination with uncovering fragments of history frozen in time.

Having relocated to Middle Tennessee to begin college in 2015, Eros has fervently pursued his photography endeavors. His heart lies in capturing the essence of the past before it fades into oblivion, a pursuit that drives him to seek out forgotten relics and stories hidden within the folds of time.

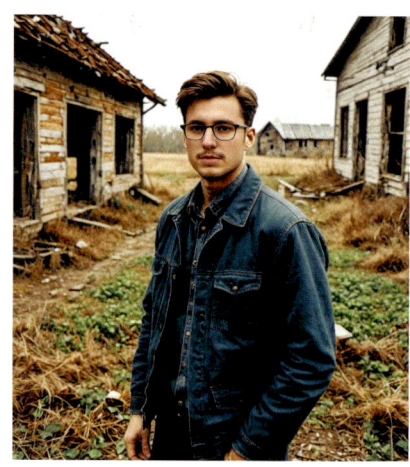